Mikail Oliveira
Breno Freitas

Population decline and the rational breeding of Bombus bees

Mikail Oliveira
Breno Freitas

Population decline and the rational breeding of Bombus bees

Changes in the body size of bees and initiatives for rational breeding of the species Bombus brevivilus

Imprint

Any brand names and product names mentioned in this book are subject to trademark, brand or patent protection and are trademarks or registered trademarks of their respective holders. The use of brand names, product names, common names, trade names, product descriptions etc. even without a particular marking in this work is in no way to be construed to mean that such names may be regarded as unrestricted in respect of trademark and brand protection legislation and could thus be used by anyone.

Cover image: www.ingimage.com

This book is a translation from the original published under ISBN 978-613-9-66073-5.

Publisher:
Sciencia Scripts
is a trademark of
Dodo Books Indian Ocean Ltd. and OmniScriptum S.R.L publishing group

120 High Road, East Finchley, London, N2 9ED, United Kingdom
Str. Armeneasca 28/1, office 1, Chisinau MD-2012, Republic of Moldova, Europe
Printed at: see last page
ISBN: 978-620-7-98513-5

"If you cry because you haven't seen the sunset, your tears won't let you see the stars"

Bob Marley

ACKNOWLEDGEMENTS

To my parents, my sister and all my family for their support and affection and for giving me everything I've ever needed in my life.

To my wife Isabela Maia da Costa for her love and for always being by my side.

To the Federal University of Ceará (UFC) and the Integrated Doctorate Programme in Zootechnics (PDIZ).

To CAPES, for the grant awarded during my doctoral studies, and to the Science Without Borders programme for the investment during my internship abroad. To my advisor Professor Dr Breno Magalhães Freitas for his participation during my academic training and for his valuable support since I was an undergraduate.

To my friend, Dr Marcelo Casimiro Cavalcante, for his participation and help in field activities and adventures with *Bombus* bees.

To the UFC Bee Research Group, especially Hiara Marques Menezes, David Silva Nogueira and Ângela Maria da Silva Gomes for their essential and valuable help and for sharing the stings with me.

To Dr Francisco Deoclécio Guerra Paulino for his essential suggestions for carrying out this work and for making the areas available for the experiments in the Bee Sector of the UFC.

Dr Favízia Freitas de Oliveira for the taxonomic identification of the species *Bombus brevivillus*.

To Professor David Kleijn, from Wageningen University, for his help during my research abroad.

To the Alterra Institute and Wageningen University, in the Netherlands, for allowing the data to be collected and for providing all the equipment needed to carry out this research abroad.

The secretary of the Postgraduate Programme in Zootechnics, Francisca das Chagas Prudêncio Bezerra.

To all the professors and staff of the Department of Animal Science at the UFC and the Ecology Research Group at Wageningen University.

SUMMARY

General Summary - Bee populations have been declining all over the world, and large bees, because they require larger supplies of food, are being more affected by environmental changes. This creates the need to develop alternatives and conservation programmes for these bees. Changes in body size are indicative of this population decline. The aim of this research was to verify changes in the body size of bees, especially large bees such as the genus *Bombus*, and to develop rational breeding initiatives for the conservation of the native species *Bombus brevivillus*, generating new information about this species, which has been little studied in the Brazilian Northeast. Of the thousands of specimens in the Naturalis Museum in the Netherlands, we measured 4510, divided into 18 different species, approximately 250 individuals per species, between males and females and distributed between the years 1866 and 2013. We also described the nesting behaviour and colony characteristics of *B. brevivillus* in the state of Ceará, Brazil, and transferred four colonies to wooden boxes that allowed for rational breeding. We obtained information on the pollen types used, the foraging pattern, the collection of resources, the influence of climatic variations on the external activities of the workers and on the development and establishment of the colonies in rational boxes. The results suggest that 18 species of bees studied are becoming smaller over time, with this reduction in size being greater in large females. With regard to the native bee *B. brevivillus,* the observations showed that this species appears to be opportunistic with regard to nesting sites and protection of offspring and does not invest much in nest building and offspring protection. The colonies produced an excellent number of workers, who, even in rational boxes, foraged intensively and completed their cycle normally, characteristics that could be useful for commercial pollination purposes.

Keywords: Pollinator decline, colony size, *Bombus* breeding, large bees.

INITIAL CONSIDERATIONS

Bees are the main pollinators of wild and cultivated plants around the world and are responsible for maintaining plant biodiversity and producing much of the food we humans consume. However, especially since the second half of the 20th century, most bee species have suffered a huge population decline, and among the main causes of this decline are changes in land use, the intensification of farming, deforestation, fires, the intense use of pesticides and climate change.

Of all the species, those with larger body sizes may be more affected and suffer more from the decline than smaller bees. This is due to the fact that these bees need larger supplies of food than smaller bees. Several studies around the world suggest this and point to the species of the genus *Bombus,* known for their large size, as one of those that suffer most from all these environmental changes.

Bombus bees are widely used in pollination programmes for agricultural crops, mainly in closed environments (vegetation houses) and some companies, most of them European, produce colonies in laboratories and make millions every year from selling and exporting them to farmers in various countries around the world, but not to Brazil, which prohibits this type of import. It would therefore be interesting to find out whether large bees are really being impacted more than smaller ones, and if this is true, to learn as quickly as possible how to work with *Bombus* species native to Brazilian territory.

There are six identified species in Brazil and among them, the species *Bombus brevivillus* has never been studied and there is very little information about it in the literature. This work therefore sets out to: firstly, investigate whether large bees do in fact suffer a greater impact and secondly, whether there is a possibility of generating information about the native species *B. brevivillus* with the aim of breeding it rationally as a way of contributing to its zootechnical exploitation and consequent conservation of the species.

As Brazil does not have a historical series of bee collections (large entomological collections), we searched for data in Europe on the influence of environmental and climatic changes on large bees, and based on this information we worked with the native Brazilian species *Bombus brevivillus.*

CHAPTER 1

Theoretical Framework

THE DECLINE IN BODY SIZE OF LARGE BEES: RATIONAL BREEDING INITIATIVES FOR THE CONSERVATION OF THE NATIVE SPECIES *Bombus (Thoracobombus) brevivillus*

1.1. THE IMPORTANCE OF BEES IN POLLINATION

Pollination is one of the most important processes in maintaining the diversity and abundance of most flowering plant species (Kevan & Viana, 2003; Klein *et al.,* 2007), and is a fundamentally important factor in the management of many agricultural crops around the world. Well-performed pollination promotes various improvements in the agricultural production chain, such as an increase in the number of pods or fruit set, an increase in the number of grains per pod, the production of better-shaped fruit of superior quality (Williams *et al.,* 1991), consequently increasing the producer's profitability.

Plants need pollinating agents to reproduce (Klein *et al.,* 2007). Pollination by animals (biotic pollination) is the most important (Keams *et al.,* 1998) and contributes to improving the quality and/or quantity of fruit and seeds (Roubik, 1995). Bees are the most important pollinating animals and participate in the reproduction of most cultivated plants (Couto and Couto, 2002). According to data from the FAO *(Food and Agriculture Organisation),* it is estimated that approximately 73% of cultivated plant species in the world are pollinated by some species of bee (FAO, 2004) and that 33% of human food depends on bees (Klein *et al.,* 2007).

The *Apis mellifera* species has characteristics such as a generalist feeding habit, a large number of individuals in the same colony, great ability to recruit several broods for foraging, perfect manageability due to the use of standardised hives, known biology and great intensity in collecting resources (Hogendoom, 2004; Morais *et al.,* 2012; Winston, 2003), and is widely used in agricultural pollination. The presence of these bees leads to gains in production and can interfere with the productive characteristics of fruit and seeds of economic interest, such as their mass, shape and size (Free, 1993).

A. mellifera bees are potential pollinators of various agricultural crops, but they are not always the most suitable for a particular crop (Banda & Paxton, 1991). In addition, *Apis mellifera* workers don't adapt very well to being kept indoors, due to the stress of confinement and temperature fluctuations (Free, 1993; Guerra Sanz, 2008). What's more, their extremely defensive behaviour, the presence of a stinger and their organisation in attacking possible enemies near the hive, makes it difficult to carry out cultivation and work inside the greenhouse, causing inconvenience to producers and workers (Amano, 2004; Cruz & Campos, 2009).

An alternative for agricultural pollination is the use of native stingless bees from the *Meliponini* and *Trigonini* tribes. Some species can be kept in rational boxes (Fabichak, 1989), which facilitates targeted management for pollination in protected environments and open fields. In addition, these bees do not have a functional sting, have a level of social organisation comparable to that of honey bees, have low defensibility, a smaller foraging flight range, colonies are perennial and adapt very well to confined conditions (Malagodi-braga *et al.,* 2004; Cruz *et al.,* 2005).

Another option for agricultural pollination is the use of solitary bees. Approximately 85% of the more than 20,400 known bee species in the world (Catalogue of life, 2015) are solitary, and these are mainly responsible for the natural pollination that occurs in wild and agricultural crops (hnperatriz-Fonseca *et al.,* 2012). Brazil has a great diversity of solitary and para-social bee species from different genera, such as *Xylocopa, Centris, Megachili* and *Halictidi* and six species from the genus *Bombus* (Moure & Melo, 2012). All the species present here certainly have potential importance as specialised pollinators (Imperatriz-Fonseca & Dias, 2004) but, especially for the genus *Bombus,* more research is still needed, as there is little information on the role of these bees in the pollination of agricultural crops.

In Europe and North America, solitary bees and *Bombus* bees have been recognised for their role in agriculture for over 40 years, promoting significant increases in the final productivity of various agricultural crops (Alves-dos-Santos, 2004; Velthuis & Doom, 2006). Examples around the world include the use of the species *Megachile rotundata* to pollinate alfalfa *(Medicago sativd),* as it is one of the most successful models for the large-scale use of wild bees for pollination (Richards, 2001; Pitts-Singer & Cane, 2011). In Europe, species from the genus *Osmia* are used to pollinate almonds *(Prunus dulcis),* apples *(Malus domestica'),* pears *(Pyrus communis),* raspberries *(Rubus* sp.), strawberries
{Fragaria x ananassd) and apricot *{Prunus armenicã)* (Felicioli *et al.,* 2004; Bosch & Kemp, 2002). These bees are also used on a large scale to pollinate various agricultural crops in the USA, Canada and Japan (Sekita, 2001; Bosch *et al.,* 2000; Torchio, 1985).

In Brazil, the importance of pollination by bees has been recognised for some years (Couto & Couto, 2007; Imperatriz-Fonseca *et al.,* 2012). Several studies have already reported the importance of solitary bees in the pollination of agricultural crops, but there are still few studies evaluating the productive effects of the introduction and management of rational farms, which are extremely important for their conservation. Magalhães & Freitas (2013), through the introduction of artificial nests for *Centris* bees in acerola *(Malpighia emarginata)* growing areas, found increases in the fruit production of this crop, while Freitas & Oliveira-Filho (2001*)*, through the introduction of artificial nests for *Centris* bees in acerola (*Malpighia emarginata)* growing areas, found increases in the fruit production of this crop. Oliveira-Filho (2001), by introducing artificial nests of *Xylocopa frontalis,* a large species popularly known as mamangava, observed a 92 per cent increase in the production of yellow passion fruit *{Passifora edulis).*

With regard to native stingless bees from the *Meliponini* and *Trigonini* tribes, various studies have shown that stingless bees are effective pollinators (Malagodi-Braga & Kleinert, 2004; Cruz *et al.,* 2005; Del Sarto *et al.,* 2005). Bees of the *Melipona quadrifasciata* species have been found to be efficient pollinators of tomato plants in protected environments, resulting in an increase in fruit quality and a reduction in mechanical damage when compared to traditional manual pollination (Del Sarto *et al.,* 2005). The jandaíra bee *{Melipona subnitida)* efficiently pollinates chilli flowers *{Capsicum annuum),* with the crop showing heavier, wider fruit with a greater number of seeds and a lower proportion of deformations. In addition, the jandaíra bee has adapted well to use in the greenhouse, having quickly started visiting the flowers (Cruz *et al.,* 2005).

Other bees widely used in the pollination of agricultural crops are the primitive solitary bees of the genus *Bombus,* where some species are produced in laboratories by specialised companies and sold to farmers in various countries who use the colonies mainly inside vegetation houses (Velthuis & Doom, 2006).

1.2.   bees of the genus *bombus* and their use in pollination of agricultural crops

Currently, few bee species are successfully bred for use in the pollination of agricultural crops, among which we can highlight the *Bombus* bees. Bees of this genus are widely used to pollinate crops in protected environments and there is a great demand for them to be bred on a large commercial scale for this purpose (Velthuis & Doom, 2006), especially the *Bombus terrestris* and *Bombus impatiens* species. A large amount of research on *Bombus* bees shows their efficiency in pollination in protected environments and the increase in characteristics of agricultural interest that is caused in fruit and seeds as a result of the rational use of these bees (Shipp *et aL,* 1994; Meisels & Chiasson, 1997; Daçgan & Ozdogan, 2004; Vergara & Fonseca-Buendía, 2012).

The rational breeding of *Bombus* bees began with studies carried out by Sladen (1912), who developed technologies for starting colonies using fertilised queens collected in the field. By 1987, a Dutch biotechnology company already had the knowledge and technology needed to produce colonies on a large scale for commercial use. It was from then on that the use of *Bombus* colonies, especially *B. terrestris,* became more widespread among producers around the world (Velthuis & Doom, 2006).

After a few years, large multinational biotechnology companies were and still are responsible for commercialising thousands of colonies every year (Velthuis & Doom, 2006). The use of colonies of bees of the genus *Bombus* generates an improvement in agricultural productivity and product quality, promoting greater food safety and other benefits for the end consumer, since farmers who use bees to pollinate their agricultural crops have cut down on pesticides and started using biological control methods to combat pests and diseases (Sande, 1990; Banda & Paxton, 1991; Presman *et al.*, 1999).

Today, *Bombus* bees are highly valued for the pollination services they provide (Potts *et* al., 2015). Five different species are sold worldwide *(B. terrestris, B. lucorum, B. occidentalis, B. ignitus and B. impatiens)* and each colony is sold for an average of 60 US dollars and used in countries in Europe, North Africa, Asia and South America (Velthuis & Doom, 2006). There are many studies that prove the pollinating efficiency of different species of *Bombus* and the increases in production caused by the use of these bees. In Mexico, with the tomato crop, the fruit produced from flowers pollinated by *Bombus ephippiatus* bees showed a greater quantity of seeds, higher total sugar content and greater fresh fruit weight at the end of the production cycle (Vergara & Fonseca-Buendía, 2012). With the same crop, but now in the USA, the *B. vosnesenskii* species caused a large increase in productivity and flowers pollinated by bees had larger fruit when compared to those that were not visited, the authors also mention the great efficiency of this species in a protected environment (Dogterom *et aL,* 1998). Another study using *B. impatiens* bees found that for the pepper crop *(Capscicum annuum),* certain production characteristics such as fruit weight, width and volume and seed weight were significantly higher in the treatment using pollination by these bees. The fruit produced was of better quality and the time taken to harvest it was shorter (Shipp *et aL,* 1994). Meisels & Chiasson (1997) found that just three workers of this species of bee are enough to effectively pollinate around 425 plants and that approximately 176 bees would be able to efficiently pollinate a hectare containing 25,000 pepper plants *(Capscicum annuum).*

The species most used in agricultural pollination, *Bombus terrestris,* was positively associated with the minimum exportable weight of melons, with around 90% of the flowers that received visits from bees producing fruit that reached the minimum exportable weight more quickly (Fisher & Pomeroy, 1989). Dasgan & Ozdogan (2004), in New Zealand, found that the vibration behaviour *("Buzz pollination")* of these bees on tomato flowers in the greenhouse caused an increase of around 90% in final production when compared to manual vibration.

Bees of the genus *Bombus* have long been used to pollinate agricultural crops. These bees contribute to increasing productivity in agricultural areas, but their importance in today's agriculture is much greater when it comes to protected cultivation (Velthuis & Doom, 2006).

Bombus bees are better adapted to green houses than other species (Free, 1993), and due to their larger size they can carry larger loads of pollen and visit several flowers before returning to the colony. They are very fast bees, visiting twice as many flowers per minute as most other species and are much more efficient pollinators compared to smaller bees (Heinrich, 2000). Around 95% of their use is in greenhouses (Velthuis & Doom, 2006), with yield increases in tomatoes *(Solanum lycopersicum)*, for example, which can exceed 28%, as well as obtaining higher quality fruit (Sande, 1990; Fiume & Parisi, 1994). However, the use of *Bombus* in pollination only became a reality once colonies began to be produced on a commercial scale.

At the beginning of the 1990s, the use of protected cultivation in Brazil began to be widely used (Grande *et al.,* 2003). With this new form of cultivation, the use of *Bombus* bees would be very interesting for improving the productivity of our planted crops, as is the case in much of Europe. However, the import of exotic bee species can have severe negative effects, such as competition with native pollinators for floral resources, competition for nesting sites, the introduction of diseases and pathogens that can infect native organisms (Goulson, 2003) and is therefore prohibited by Brazilian legislation (Saraiva *et al.,* 2012).

1.3.   DECLINE IN BEE POPULATIONS

Despite all the importance mentioned, bees are in the process of disappearing in some parts of the world, mainly in Europe and some countries in North America (Potts *et al.,* 2015). Recent research is showing a major decline in native bees and honey bees *(Apis melliferá)* in several countries around the world (Kleijn & Raemakers, 2008; Freitas *et al.,* 2009; Biesmeijer *et al.,* 2006). Environmental changes, especially as a result of agricultural intensification, are causing severe negative effects on the availability of food, making it less abundant, less predictable and present for shorter periods of time, directly impacting bees that depend on floral resources. Added to this is the introduction of exotic species, deforestation, burning, ecosystem fragmentation, the action of sharecroppers (extractivism), heavy application of pesticides and improper land use (Lopes *et al.* 2005; Kerr, *et al.,* 2005).

Apis mellifera, the honey-producing species that is widely used in agricultural pollination, is suffering from declining populations. CCD (Colony Collapse Disorder) has been affecting colonies all over the world and possibly even in certain areas of Brazil, causing a great deal of damage to agriculture, including a significant reduction in food production. This disorder has already caused serious drops in the number of *A. mellifera* colonies in some countries (Aizen & Harder, 2009; Engelsdorp *et al.,* 2008) and particularly in the United States, where according to the National Agriculture Statistics Service (2008) a systematic loss of around 30 per cent of colonies per year is occurring.

Concern about pollination deficits and the risk of depending almost exclusively on a single species as a pollinator, especially in Europe and North America, has led to initiatives aimed at conserving pollinators, favouring many research studies on native species (Dias *et al.,* 1999; Imperatriz-Fonseca *et al.,* 2006). In the United Kingdom, the abundance of solitary bees decreased by around 52 per cent in previous years (Biesmeijer *et al.,* 2006). In the Centre-West region of Brazil, the actions of extractivists in search of honey from native stingless bees alone have drastically reduced the abundance of four different species of these bees, *Melipona rufiventris, Melipona bicolor, Melipona marginata* and *Cephalotrigona femorata* (Kerr, *et al.,* 2005).

The discussion on the loss of pollinators and their pollination services is based on recent evidence of their decline on a local and regional scale (Biesmeijer *et al.,* 2006) and the high rate of extinction in various groups (Millennium Ecosystem Assessment, 2005), in addition to the record of significant losses in managed pollinator populations (Allen-Wardell *et al.,* 1998). To help raise awareness and decision-making in environmental and social policies, economic values have been estimated for pollination services, verifying amounts in the billions (Gallai *et al.,* 2009). Therefore, reductions in the populations of bees that pollinate agricultural crops can cause serious economic damage and may mean that the activity becomes unviable.

1.4.   the decline of bees of the genus *bombus*

It is widely accepted by the scientific community that several species of bees in the genus *Bombus* have disappeared and that others are clearly at risk. The abundance and diversity of these bees are now declining (Biesmeijer *et al.,* 2006; Olroyd, 2007, Potts *et al*, 2015). Changes in land use, especially after the second half of the 20th century, have caused severe habitat loss and created important limiting factors for bee populations, especially for bees of the genus *Bombus*, which are large bees and require more floral resources throughout their lives (Goulson, 2003; FAO, 2006).

Bombus bees, popularly known as ground bees, are vitally important for the natural pollination of wild plants and are also widely used in agricultural pollination (Velthuis & Doom, 2006). However, these bees are disappearing very quickly, and there is a consensus that the decline in the number and diversity of *Bombus* bees is linked to the intensification of livestock farming and agricultural practices (Williams, 1986; Osbome & Corbet, 1994; Goulson, 2003).

In much of Europe, some species of *Bombus* have become extinct and many others are declining dramatically (Biesmeijer *et al.* 2006), especially after 1950. During the second half of the 20th century (1951 - 2000), four species of this genus became extinct in eleven European countries *(B. armeniacus, B. cullumanus, B. serrisquama* and *B. sidemii)* (Kosior *et al*, 2007), in the United Kingdom three of twenty-five species of *Bombus* became extinct, and eight are already showing a major decline in abundance (UK Biodiversity Group, 1998; Goulson, 2003).

In the Netherlands, Kosior *et al.* (2007) found that the population size of six species is of little concern *(Bombus hortorum, B. hypnorum, B. lapidarius, B. pascuorum, B. pratorum* and *B. terrestris),* but the same research described the extinction of four species *(Bombus confusus, B. pomorum, B. cullumanus* and *B. sylvarum), and warned of three critically* endangered species *(Bombus distinguendus, B. ruderatus, B. subterraneus). pomorum, B. cullumanus* and *B. sylvarum),* and warned of three species that are critically endangered *(Bombus distinguendus, B. ruderatus* and *B. subterraneus)* and six species that are at risk of extinction *(Bombus soroeensis, B. veteranus, Bombus humilis, B. jonellus, B. muscorum* and *B. ruderarius).*

The consequences of intensive land use are not a new topic, but a very important one. It affects the availability of food, nectar and pollen, and the preference for a particular plant species, which have been suggested as key factors associated with the decline of *Bombus* bees (Rasmont, 1988, Goulson & Darvill 2004, Goulson *et al.*, 2005, Kleijn & Raemakers, 2008). The bees of this genus in Brazil are probably suffering these same impacts, but studies on the species of *Bombus* that occur in Brazilian territory are still very scarce, and there is very little about the species *Bombus brevivillus,* for example. Added to this is the fact that the ground-dwelling mamangava bees that occur in Brazil behave in such a way that the nests of these native species are constantly destroyed by the population (Garófalo, 1980).

The worldwide importance of these bees in the pollination of agricultural crops is undeniable and, in order to avoid importing exotic species, an idea that certainly has many supporters, we need to study the species that exist here, because surely among them we will find competent pollinators for many of our crops.

1.5. PERSPECTIVES FOR THE USE OF NATIVE BEES OF THE GENUS *Bombus* IN AGRICULTURAL POLLINISATION PROGRAMMES

With the growth of this form of agriculture indoors and the rise of this type of cultivation in Brazil, the use of *Bombus* bees to pollinate vegetables indoors would be an excellent tool for increasing productivity. However, the importation of colonies for this purpose is only viewed favourably by farmers who are looking for the possibility of using these exotic bees in their crops, but the introduction of exotic species can cause severe environmental impacts and the Brazilian government prohibits this commercial practice (Goulson, 2003; Saraiva *et al.*, 2012).

There are six species of this genus in Brazil, all belonging to the subgenus Thoracobombus: *Bombus atratus* Franklin, 1913, *B. bellicosus* Smith, 1879, *B. brasiliensis* Lepeletier, 1836, *B. brevivillus* Franklin, 1913, *B. morio* Swederus, 1787 and *B. transversalis* Olivier, 1789 (Moure & Sakagami, 1962), and there is no doubt that efficient pollinators can be found among them for many of Brazil's agricultural crops.

The fact that their nests are constantly destroyed (Garófalo, 1980) makes it increasingly difficult to find them in the forest and, together with the very aggressive behaviour of these bees, it is very difficult to carry out research that contributes to the use of these native bees in the pollination of agricultural crops. Benavides (2008) studied the establishment of *Bombus morio* colonies under captive conditions and through laboratory crosses, it was possible to induce the oviposition of queens, but the continuation and full development of these colonies was not achieved.Research into native bees of the genus *Bombus* is still very scarce, the species *B. atratus* and *B. morio* have been the most studied to date, research into *B. transversalis* and *B. brasiliensis* is still small, while practically nothing is known about *B. bellicosus* and *B. brevivillus*. It is therefore extremely important to carry out research that can contribute to obtaining additional information on these bees in order to subsequently develop management techniques that will allow these native bees to be bred and used to pollinate agricultural crops, especially in protected environments, and to introduce them into environments where their population has been decreasing. With the progress of research in this area, in the very near future, proof of the efficiency of native bees will not only contribute to more sustainable and productive agriculture, but will also favour the preservation of *Bombus* species native to Brazilian territory.

1.6. BEHAVIOUR OF BEES OF THE GENUS *Bombus* IN THE FORMATION AND DEVELOPMENT OF THE COLONY

Bees of the genus *Bombus* are a primitive eusocial group, with care for offspring and division of labour within the colony (Michener, 1974). Initially, the colony is formed from the queen (solitary life stage), during which time the queen does all the work of collecting food, building pots and cocoons and caring for the young (Goulson, 2003).

When the first worker daughters of the queen begin to emerge, the queen ceases her external work and takes on only the reproductive function within the colony, while the workers take on all the internal and external foraging work inherent in maintaining the colony (Heinrich, 2000). Males have only the reproductive function and workers, as in most eusocial bee species, can lay eggs from ovarian development, giving rise only to males, workers and potential queens can only be produced from the laying of the queen.

The moment the first males and new queens begin to emerge, a conflict begins within the colony, this moment is known as *kin-selection* or *'kin-selected conflict'* (Hamilton, 1964). Individual values between one individual and another within the colony vary according to their genetic relationships. Thus, according to these differences in genetic relationships, conflict begins between groups, where different parts are unequally related to the offspring (Trivers & Hare, 1976; Ratnieks & Reeve, 1992).

In social insects that are initiated by a founding queen, as in the genus *Bombus,* queens are more closely related to their own offspring than to the offspring of workers, while workers are more closely related to their offspring and the offspring of other workers than to the offspring of the queen.

The theory of kin selection predicts conflicts between the percentage of males, with queens favouring males produced by the queen and workers favouring the appearance of males produced by the workers themselves (Hamilton, 1964; Trivers & Hare, 1976). *Bombus terrestris* workers with activated ovaries in mature colonies try to exchange the eggs of queens and other workers for their own eggs and initiate attacks on the queen (Van Honk *et al.,* 1981; Van der Blom, 1986), studies suggest that this behaviour is an expression of conflict in kin selection (Bourke & Ratnieks, 2001).

This moment when the workers lay haploid eggs is known as*' competition point'* (Duchateau & Velthuis 1988). It is hypothesised that this happens because of the reduction in the queen's pheromone, which is detected by the workers.

From this signal onwards, the workers recognise that the queen no longer represents a future reproductive resource of full sisters, so that the laying of eggs by males is now their kinship selection of interest (Cnaani*, et al.,* 2000; Bourke & Ratnieks, 2001).

From this point onwards and after the reproductive individuals (males and new queens) begin to emerge, the colony has a high worker mortality rate, beginning the final part of the reproductive cycle for bees of this genus. With the reduction in the number of workers, food becomes scarce and, consequently, the founding queen dies. At this point, which in temperate climates coincides with the start of the winter season, the sexed workers have already abandoned the colony, which weakens considerably and becomes more susceptible to attacks from enemies and parasites, causing the colony to die out completely at the end of the reproductive cycle (Alford, 1975). However, in tropical climates, according to Sakagami *et al.* (1967) and Garófalo *et al.* (1986) after the death of the founding queen, new queens remain in the nest, and after copulation, reactivate the mother nest, as has been described for the Neotropical species *Bombus atratus.*

1.7.   BIBLIOGRAPHICAL REFERENCES

AIZEN, M. A. & HARDER, L. D. The global stock of domesticated honey bees is growing slower than agricultural demand for pollination. **Current Biology,** London, v. 19, p. 915-918,2009.

ALFORD, D.V. Bublebees. London: Davis - Poynter, 1975.

ALLEN-WARDELL, G., BERNHARDT, P.; BITNER, R.; BURQUEZ, A.; BUCHMANN, S.L.; CANE, J. H.; COX, P.A.; DALTON, V.; FEINSINGER, P.; INOUYE, D.; INGRAM, M.; JONES, C.E.; KENNEDY, K.; KEVAN, P.;

KOOPOWITZ, H.; MEDELLIN, R.; MEDELLIN-MORALES, S.; NABHAN, G.P.; PAVLIK, B.; TEPEDINO, V.J.; TORCHIO, P. & WALKER, S. The potential consequences of pollinator declines on the conservation of biodiversity and stability of food crop yields. **Conservation Biology** ,v. 12, p. 8-17, 1998.

ALVES DOS SANTOS, I. Knowledge and breeding of solitary bees: a challenge. **Revista Tecnologia e Ambiente,** v. 10, p. 99-113, 2004.

AMANO, K. **Attempts to introduce stingless bees for the pollination of crops under greenhouse conditions in Japan.** Laboratory of Apiculture, National Institute of Livestock and Grassland Science, Japan, 2004. Available at: <http://www.fftc.agnet.org/library.php?func=view&id=20110913145638>.

BANDA, H.J. & PAXTON, R.J. Pollination of greenhouse tomatoes by bees. In: **VI International Symposium on Pollination. Acta Horticulturae,** (ISHS) 288, p. 194- 198,1991.

BENAVIDES, M.L.A. **Aspects of the reproductive biology of _Bombus morio_ (SWEDERUS) and _Bombus atratus_ FRANKLIN (HYMENOPTERA, APIDAE).** Master's dissertation. Master's degree in entomology. Federal University of Viçosa, Minas Gerais. 2008.

BIESMEIJER J.C., ROBERTS, S.P.M., REEMER, M., OHLEMULLER, R., EDWARDS, M., PEETERS, T., SCHAFFERS, A.P., POTTS, S.G., KLEUKERS, R., THOMAS, C.D., SETTELE, J., KUNIN, W.E. Parallel declines in pollinators and insect-pollinated plants in Britain and The Netherlands. **Science,** v. 313, p. 351-354, 2006.

BOSCH, J. & KEMP, W.P. Developing and establishing bee species as crop pollinators: the example of _Osmia spp._ and fruit trees. **Bulletin of Entomological Research,** v. 92, p. 3-16, 2002.

BOSCH, J.; KEMP, W.P. & PETERSON, S.S. Management of _Osmia lignaria_ (Hymenoptera, Megachilidae) populations for almond pollination: methods to advance bee emergence. **Environmental Entomology,** v. 29, p. 874-883,2000.

BOURKE, A. F. G. & RATNIEKS, F. L. W. Kin-selected conflict in the bumble-bee _Bombus terrestris_ (Hymenoptera: Apidae). **Proceedings of the Royal Society of London, Series B,** v. 268, p. 347-355, 2001.

CNAANI, J., ROBINSON, G. E., BLOCH, G., BORST, D. & HEFETZ, A. The effect of queen-worker conflict on caste determination in the bumblebee *Bombus terrestris*. **Behavioural Ecology and Sociobiology,** v. 47, p. 346-352, 2000.

COUTO, R. H. N. & COUTO, L. A. **Apicultura: manejo e produtos.** 2 ed. Jaboticabal: FUNEP, 191 p., 2002.

COUTO, R.H.N. & COUTO, L.A. Use of pollinators in the conservation and sustainability of agriculture. **Mensagem Doce,** n. 90, 2007.

CRUZ, D. O.; FREITAS, B. M.; SILVA, L. A.; SILVA, E. M. S. & BOMFIM, I. A. Pollination efficiency of the stingless bee *Melipona subnitida* on greenhouse sweet pepper *(Capsicum annuum).* **Revista Agropecuária Brasileira,** v. 40, p. 1197-1201, 2005.

CRUZ, D.O. & CAMPOS, L.A.O. Pollination by bees in protected crops. **Revista Brasileira de Agrociência,** v. 15, n. 1-4, p. 5-10,2009.

DASGAN, H. Y. & OZDOGAN, A. O. Effectiveness of bumblebee pollination in anti- frost heated tomato greenhouses in the Mediterranean Basin. **Turkish Journal of Agriculture and Forestry,** v. 28, p. 73-82, 2004.

DEL SARTO, M.C.L.; PERUQUETTI, R.C. & CAMPOS, L.A.O. Evaluation of the Neotropical stingless bee *Melipona quadrifasciata* (Hymenoptera: Apidae) as pollinator of greenhouse tomatoes. **Journal of Economic Entomology,** v. 98, p. 260-266,2005.

DIAS, B.S.F.; RAW, A. & IMPERATRIZ-FONSECA, V.L. **International Pollinators Initiative: The São Paulo Declaration on Pollinators.** Report on the recommendations of the Workshop on the Conservation and Sustainable Use of Pollinators in Agriculture with Emphasis on Bees. Ministry of Environment (MMA), University of Sao Paulo (USP) and Brazilian Corporation for Agricultural Research (Embrapa), Brasília. 79p, 1999.

DOGTEROM, M. H.; MATTEONI, J. A. & PLOWRIGHT, R. C. Pollination of greenhouse tomatoes by the North American *Bombus vosnesenskii* (Hymenoptera: Apidae). **Journal of Economic Entomology,** v. 91, n. 1, p. 71-75, 1998.

DUCHATEAU, M. J. & VELTHUIS, H. H. W. Development and reproductive strategies *in Bombus terrestris* colonies. **Behaviour,** v. 197, p. 186-207, 1988.

DUCHATEAU, M. J. Regulation of colony development in bumblebees. **Acta Horticulturae,** v. 288, p. 139-143, 1991.

ENGELSDORP, D. van; HAYES, JÚNIOR.; UNDERWOOD, R. M.; PETTIS, J. S. A survey of honey bee colony losses in the U.S., Fall 2007 to Spring 2008. **PLoS ONE,** San Francisco, v. 3, n. 12, p. 40-71, 2008.

FABICHAK, I. **Jataí indigenous stingless bees.** São Paulo: Nobel, 1989.
FAO. Conservation and management of pollinators for sustainable agriculture - the International response. In: Freitas, B.M.; Pereira, J.O.P. (eds.) Solitary bees: conservation, rearing and management for pollination. University Press. Fortaleza, Brazil, p. 19-2, 2004.

FAO. **Livestock's Log Shadow.** 416p., 2006.

FELICIOLI A., KRUNIC' M. & PINZAUTIM. **Rearing and using *Osmia* bees for crop pollination: a help from a molecular approach.** p. 161-174. In: B. M. FREITAS & J.O.B. PORTELA (Eds.). *Solitary bees:* conservation, rearing and management for pollination. University Press, UFC, Fortaleza. 285p., 2004.

FISHER, R.M. & POMEROY, N. Pollination of greenhouse muskmelons by bumble bees (Hymenoptera: Apidae). **Journal of Economic Entomology,** v. 82, n. 4, p. 1061- 1066,1989.

FIUME, F. & PARISI, B. Fitoregolatori e bombidi nella frittificazione del pomodoro. **Colture Protette,** v. 10, n. 87-93,1994.

FREE, J.B. **Insect pollination of crops.** 2ª ed. London: Academic Press, 684 pp., 1993. FREITAS, B.M. & OLIVEIRA FILHO, J. H. **Rational breeding of mamangavas for pollination in agricultural areas.** Fortaleza: Banco do Nordeste. 96p, 2001.

FREITAS, B.M.; IMPERATRIZ-FONSECA, V.L.; MEDINA, L.M.; KLEINERT, A.M.P.; GALETTO, L.; NATES-PARRA, G. & QUEZADA-EUÁN, J.J.G. Diversity, threats and conservation of native bees in the Neotropics. **Apidologie,** v. 40, p. 332-346, 2009.

GALLAI, N.; SALLES, J. M.; SETTELE, J.; VAISSIÈRE. B. E. Economic valuation of the vulnerability of world agriculture confronted with pollinator decline. **Ecological Economics,** Amsterdam, v. 68, p. 810-82, 2009.

GARÓFALO, C.A. Bionomic aspects of *Bombus* (Fervidobombus) *morio* (Swederus). Worker size and colony development (Hymenoptera, Apidae). **Rev. Brasil. Biol.,** v. 40, n. 2, p. 345-348, 1980.

GOULSON, D. & DARVILL, B. Niche overlap and diet breadth in bumblebees; are rare species more specialised in their choice of flowers? **Apidologie,** v. 35, p. 55-63, 2004.

GOULSON, **D. Bumblebees: Behaviour and Ecology.** Oxford, UK: Oxford Univ. Press, 2003.

GOULSON, D., HANLEY, M.E., DARVILL, B., ELLIS, J.S. & KNIGHT, M.E. Causes of rarity in bumblebees. **Biological Conservation,** v. 122, p. 1-8, 2005.

GRANDE, L.; LUZ, J.M.Q.; MELO, B.; LANA, R.M.Q. & CARVALHO, J.O.M. The protected cultivation of vegetables in Uberlândia-MG. **Horticultura Brasileira,** v. 21, n. 2, p. 241-244, 2003.

GRETENKORD, C. **Laborzucht der dunklen Erdhummel** *Bombus terrestris* **L. (Hymenoptera: Apidae) und toxikologische Untersuchungen unter Labor- und Halbfreilandbedingungen.** Ph. D. Thesis. Institut fur Landwirtschaftliche Zoologie und Bienenkunde. Reinische Friedrich-Wilhelms-Universität, Bonn, Germany, 1996.

GUERRA SANZ, J.M. **Crop pollination in greenhouses.** In: Bee pollination in agricultural ecosystems. JAMES, R.R.; PITTS-SINGER, T.L. (eds.). New York: Oxford University Press, Inc., ch. 3, p. 27-47, 2008.

HAMILTON, W. D. The genetical evolution of social behaviour I, II. **Journal of Theoretical Biology,** v. 7, p. 1-52,1964.

HEINRICH, B. **Bumblebee Economics.** Harvard College. United States of America, 2000.

HOGENDOORN, K. **On promoting solitary bee species for use as crop pollinators in greenhouses.** In: Solitary bees: conservation, rearing and management for pollination. FREITAS, B.M.; PEREIRA, J.O.P. (eds.), Fortaleza: University Press, p. 213-221, 2004.

IMPERATRIZ-FONSECA, V. L.; CANHOS, D. A. L.; ALVES, D. A. & SARAIVA, A. M. Pollinators and pollination - A global issue. In: IMPERATRIZ-FONSECA, V. L.; CANHOS, D. A. L.; ALVES, D. A.; SARAIVA, A. M. (Org.) **Pollinators in Brazil - contribution and perspectives for biodiversity, sustainable use, conservation and environmental services,** led. São Paulo: EDUSP, p. 335-348, 2012.

IMPERATRIZ-FONSECA, V.L. & DIAS, B.F.S. Brazilian Pollinator Initiative. In: Solitary bees and their role in pollination. In: Freitas & Pereira (ed.), **Solitary bees: conservation, rearing and management for pollination.** Fortaleza, CE, p. 27-34, 2004.

IMPERATRIZ-FONSECA. V.L.; DE JONG, D.; SARAIVA. A.M.(eds.). **Bees as Pollinators in Brazil: assessing the status and suggesting the best practices.** Holos Ed., Ribeirão Preto, 114p, 2006.

KEARNS C. A.; INOUYE, D. W. & WASER, N. M. Endangered mutualisms: the conservation of plant-pollinator interactions. **Annual Review of Ecology, Evolution and Systematics,** v. 29, p. 83-112,1998.

KERR, W.E.; CARVALHO, G.A.; SILVA, A.C. & ASSIS, M.G.P. Little-mentioned aspects of Amazonian biodiversity. **Mensagem doce,** v. 12, n. 80, 2005.

KEVAN, P.G. & VIANA, B.F. The global decline of pollination services. **Biodiversity,** v.4, n. 4, p. 3-8, 2003.

KLEIJN, D. & RAEMAKERS, A. A retrospective analysis of pollen host plant use by stable and declining bumblebee species. **Ecology,** v. 89, n. 7, p. 1811-1823, by the Ecological Society of America, 2008.

KLEIN, A.M.; VAISSIERE, B.E.; CANE, J.H.; STEFFAN-DEWENTER, L; CUNNINGHAM, S.A.; KREMEN, C. & TSCHARNTKE, T. Importance of pollinators in changing landscapes for world crops. **Proceedings of the Royal Society B: Biological Sciences,** v. 274, p. 303-313, 2007.

KOSIOR A, CELARY W, OLEJNIKZAK P, FIJAL J & KROL W, *et al.* The decline of the bumble bees and cuckoo bees (Hymenoptera: Apidae: Bombini) of western and central Europe. **Oryx,** v. 41, p. 79-88, 2007.

LOPES, M.; FERREIRA, J. B. & SANTOS, G. Stingless bees: invisible biodiversity. **Agriculturas,** v.2, n.4, 2005.

MAGALHÃES, C.B. & FREITAS, B.M. Introducing nests of the oil-collecting bee *Centris analis* (Hymenoptera: Apidae: Centridini) for pollination of acerola *(Malpighia emarginatá)* increases yield. **Apidologie,** v. 44, p. 234-239, 2013.

MALAGODI-BRAGA, K.S. & KLEINERT, A.M.P. Could *Tetragonisca angustula* (Apinae, Meliponini) be effective as strawberry pollintor in greenhouse. **Australian Journal of Agricultural Research,** v. 55, n. 7, p. 771-773, 2004.

MEISELS, S. & CHIASSON, H. Effectiveness of *Bombus impatiens* Cr. as pollinators of greenhouse sweet peppers *(Capsicum annuum* L.). **Acta Horticulturae,** v. 437, p. 425-429, 1997.

MICHENER, C.D. **The social behaviour of the bees.** The Belknap Press of Harvard University Press, 1974.

MILLENIUM ECOSYSTEM ASSESMENT. **Ecosystem and human well-being: synthesis.** Island Press, Washington, DC, lOOp, 2005.

MORAIS, M.M.; DE JONG, D.; MESSAGE, D.; GONÇALVES, L.S. Perspectives and challenges for the use of *Apis mellifera* bees as pollinators in Brazil. In: **Pollinators in Brazil: contribution and perspectives for biodiversity, sustainable use, conservation and environmental services.** IMPERATRIZ-FONSECA, V.L.;
CANHOS, D.A.L.; ALVES, D.A.; SARAIVA, A.M. (eds.), São Paulo: Edusp, chap. 10, p. 203-212, 2012.

MOURE, J.S. & SAKAGAMI, S.F. The social mamangabas of Brazil *(Bombus Latr.)* (Hymenoptera, Apidae*).* **Studia Entomologica,** v. 5, p. 65-194, 1962.

NATIONAL AGRICULTURAL STATISTICS SERVICE. Honey. 2008. Available at: <www.nass. usda.gov>. Accessed on: 10 Apr. 2010.

NOGUEIRA-NETO, P. **Vida e criação das abelhas indígenas sem ferrão.** São Paulo, SP: Ed. Nogueirapis. 442p., 1997.

OLROYD, B.P. What's killing American honey bee? **PLoSBiology,** v. 5, n. 6, p. 168, 2007.

OSBORNE, J.L. & CORBET, S.A. Managing habitats for pollinators in farmland. *Aspects of Applied Biology,* v. 40, p. 207-215, 1994.

PITTS-SINGER, T.L., & CANE J.H. The alfalfa leafcutting bee, *Megachile rotundatcr.* The world's most intensively managed solitary bee. **Annual Review of Entomology,** n. 56, v. 1,p. 221 - 237, 2011.

POTTS, S.; BIESMEIJER, K.; BOMMARCO, R.; BREEZE, T.; CARVALHEIRO, L.; FRANZEN, M.; GONZALEZ-VARO, J.P.; HOLZSCHUH, A.; KLEIJN, D.; KLEIN, A.M.; KUNIN, B.; LECOCQ, T.; LUNDIN, O.; MICHEZ, D.; NEUMANN, P.; NIETO, A.; PENEV, L.; RASMONT, P.; RATAMAKI, O.; RIEDINGER, V.; ROBERTS, S.P.M.; RUNDLOF, M.; SCHEPER, J.; SORENSEN, P.; STEFFAN- DEWENTER, I.; STOEV, P.; VILA, M.; SCHWEIGER, O. **Status and trends of European pollinators.** Key findings of the STEP project. Pensoft Publishers, Sofia, 72 p. 2015.

PRESMAN, E.; SHAKED, R.; ROSENFELD, K. & HETETZ, A. A comparative study of the efficiency of bumblebees and an electric bee in pollinating unheated greenhouse tomatoes. **Journal of Horticultural Science and Biotechnology,** v. 74, p. 101-104, 1999.

RASMONT, P. **Monographie écologique et zoogéographique des bourdons de France et de Belgique (Hymenoptera, Apidae, Bombinae).** Dissertation, Faculty of Agronomic Sciences, Gembloux, Belgium, 1988.

RATNIEKS, F. L. W. & REEVE, H. K. Conflict in single-queen Hymenopteran societies: the structure of conflict and processes that reduce conflict in advanced eusocial species. **Journal of Theoretical Biology,** v. 158, p. 33-65,1992.

RICHARDS, A. J. Does low biodiversity resulting from modem agricultural practice affect crop pollination and yield? **Annals of Botany, Oxford,** v. 88, p. 165-172, 2001.

ROSKOV, Y.; ABUCAY, L.; ORRELL, T.; NICOLSON, D.; KUNZE, T.; CULHAM, A.; BAILLY, N.; KIRK, P.; BOURGOIN, T.; DEWALT, R.E.; DECOCK, W.; DE WEVER, A., eds. Species 2000 & ITIS **Catalogue of Life,** 2015 Annual Checklist. Digital search at:

www.catalogueoflife.org/annual-checklist/2015. Species 2000: Naturalis, Leiden, the Netherlands. 2015.

ROUBIK, D. W. **Pollination of cultivated plants in the tropics.** Rome: FAO. Agricultural Services Bulletin, p. 118, 1995.

SAKAGAMI, S.F.; AKAHIRA, Y. & ZUCCHI, R. Nest architecture and brood development in a neotropical bumblebee *Bombus atratus*. **Insectes Society,** v. 14, p. 389-414, 1967.

SANDE, J. Bumblebees are a good alternative to truss vibration for beefsteak tomatoes. **Horticultural Abstracts,** v. 60, p. 506, 1990.

SANTOS, S.A.B.; ROSELINO, A.C. & BEGO, L.R. Pollination of Cucumber, *Cucumis sativus* L. (Cucurbitales: Cucurbitaceae), by the Stingless Bees *Scaptotrigona* aff. *depilis* Moure and *Nannotrigona testaceicomis* Lepeletier (Hymenoptera: Meliponini) in Greenhouses. **Neotropical Entomology,** v. 37, n. 5, p. 506-512,2008.

SARAIVA, A.M.; ACOSTA, A.L.; GIANNINI, T.C.; IMPERATRIZ-FONSECA, V.L; MARCO JÚNIOR, P. *Bombus terrestris* in South America: Possible invasion routes of this exotic pollinator to Brazil. In: **Pollinators in Brazil: contributions and perspectives for biodiversity, sustainable use, conservation and environmental services.** IMPERATRIZ-FONSECA, V.L.; CANHOS, D.A.L.; ALVES, D.A.;
SARAIVA, A.M. (eds.), São Paulo: Edusp, ch. 10, p. 203-212, 2012.

SEKITA, N. Managing *Osmia comifrons* to pollinate apples in Aomori Prefecture, Japan. **Acta Horticulture,** v. 561, p. 303-308, 2001.

SHIPP, J.L.; WHITFIELD, G.H. &PAPADOPOULOS, A.P. Effectiveness of the bumble bee, *Bombus impatiens* Cr. (Hymenoptera: Apidae), as a pollinator of greenhouse sweetpepper. **Scientia Horticulturae,** v. 57, p. 29-39, 1994.

SLADEN, F.W.L. **The Bumblebee. Its Life History and How to Domesticate It.** London: Macmillan. 1912.

TORCHIO, P.F. Fields experiments with the pollinator species *Osmia lignaria propinqua* Cresson, in apple orchard: V, (1979-1980), Methods of introducing bees, nesting success, seeds counts, fruit yields (Hymenoptera, Megachilidae). **Entomological Society,** v. 58, p. 448-464, 1985.

TRIVERS, R. L. & HARE, H. Haplodiploidy and the evolution of the social insects. **Science,** v.191, p. 249-263,1976.

UK Biodiversity Group. UK Biodiversity Group Tranche 2 Action Plans: Invertebrates, vol. IV. English Nature, Peterborough, 1998.

VAN DEN EIJNDE, J.; DE RUUTER, A.; VAN DER STEEN, J. Method for rearing *Bombus terrestris* continuously and the production of bumble bee colonies for pollination purposes. **Acta Hortic,** v. 288, p. 154-158,1991.
VAN DER BLOM, J. Reproductive dominance within colonies of *Bombus terrestris* (L.). **Behaviour,** v. 97, p. 37-49, 1986.

VAN HONK, C. J. G., RO"SELER, P.-F., VELTHUIS, H. H. W. & HOOGEVEEN, J. C. Factors influencing the egg-laying of workers in a captive *Bombus terrestris* colony. **Behavioral Ecology and Sociobiology,** v. 9, p. 9-14, 1981.

VELTHUIS, H.H.W. & DOORN, A. van. A century of advances in bumblebee domestication and the economic and environmental aspects of its commercialisation for pollination. **Apidologie,** v. 37, p. 421-451, 2006.

VENTURIERI, G.C.; RAIOL, V.F.O. & PEREIRA, C.A.B. Evaluation of the introduction of rational breeding of *Melipona fasciculata* (Apidae: Meliponina) among family farmers in Bragança, PA, Brazil. **Biota Neotropica,** v. 3, n. 2, p. 1-7, 2003.

VERGARA, C.H. & FONSECA-BUENDÍA, P. Pollination of greenhouse tomatoes by the Mexican bumblebee *Bombus ephippiatus* (Hymenoptera: Apidae). **Journal of Pollination Ecology,** v. 7, n. 4, p. 27-30, 2012.

WILLIAMS, I.H.; CORBET, S.A. & OSBORNE, J.L. Beekeeping, wildbees, wildbees and pollination in the European Community. **Bee World,** v. 72, p. 170-180,1991.

WILLIAMS, P.H. Environmental change and the distribution of British bumble bees (Bombus Latr.). **Bee World,** v. 67, p. 50-61, 1986.

WINSTON, M.L. **The biology of the bee.** Translation by Carlos A. Osowski - Porto Alegre: Magister, 276 p., 2003.

CHAPTER 2

Decrease in body size as an indication of bee population decline

Abstract - Many native bees are disappearing around the world. Changes in land use over the last century have drastically reduced the sources of natural resources available to bees. Large-bodied species have higher pollen and nectar requirements for maintenance and reproduction than smaller-bodied species, and are therefore particularly affected by all these changes in their natural habitat. Environmental pressure, such as a reduction in the availability of viable food, can force morphological adaptations in bees, especially in their body size. I therefore put forward the hypothesis that large bee species have become smaller over the last century, but smaller bee species have not. We tested this hypothesis by measuring the body size (inter-tegular distance - ITD) of 4,510 bee specimens divided into 18 different species and collected in the Netherlands over a period of 149 years. We used a Linear Mixed Model (LMMs) to examine whether body size reduced more strongly in large bees than in small bee species. We also analysed whether these possible trends in bee body size were influenced by temperature changes over all these years. The results showed that since 1866 the 18 bee species investigated have shown a decline in body size of 4.9 per cent for both sexes. The body size of females declined significantly during the approximately 150 years of the study period, and large females showed a more pronounced body size decline than small females. Between 1866 and 2013, the predicted decline in DIT was 2.3 per cent and 7.8 per cent for females with Initial Body Size of 1.7 and 6.0 mm respectively. Males similarly showed a decline in body size of 2.7 per cent since 1866, with similar trends in body size for large and small males. Temperature had no measurable relationship with bee body size.

Keywords: Body size of bees, climate change, large bees, *Bombus*.

2.1 INTRODUCTION

Bees are functionally important in natural ecosystems and agricultural areas. This very important group of animals is responsible for plant biodiversity and food production through its pollination services (Kevan & Philips, 2001; Klein *et al.,* 2007; Ollerton *et al.,* 2011), but both native bees and honey bees *(Apis melliferà)* are now declining. Some species are becoming extinct and many others are at great risk of extinction (Biesmeijer *et al.,* 2006; Olroyd, 2007), and this is inspiring a lot of research around the world to examine both the causes and consequences of this decline. As a result, evidence has been accumulating that the reduction in bee populations is directly linked to various factors, such as the destruction of their natural habitat, the intensification of farming, the indiscriminate use of pesticides, the action of pathogens and also due to climate change over the years (Potts *et al.,* 2010).

It is also becoming significantly clear that the local and regional decline of this important group of pollinators has negative consequences for the pollination services they provide (Kremen *et al.,* 2002; Bommarco *et al.,* 2012; Deguines *et al.,* 2014). Little is known about how bees adapt phenotypically or evolutionarily to these changes in their natural environment. Some individuals, through modifications in their behaviour, morphology or physiology, manage to adapt to the negative consequences of changes in environmental conditions (Hughes, 2000), and recent studies show that changes in environmental conditions can lead to heritable genetic changes in animal populations in a short space of time (Bradshaw & Holzapfel 2006; Agrawal *et al.,* 2012).

In bees, body size may be one of the first characteristics to be affected by these environmental changes. The body size of bees is closely linked to their flight range and mobility during foraging (Greenleaf *et al.,* 2007) and, consequently, to the access to floral resources that each bee needs for itself and its offspring. In this way, the body size of the offspring depends on the amount of pollen and nectar that each mother bee provides, where better provisioning will result in larger offspring (Klostermeyer *et al.,* 1973; Bosch & Vicens, 2002).

However, larger bees also need larger quantities of pollen and nectar for their maintenance and reproduction (Muller *et al.,* 2006). When the viable floral resource becomes progressively scarce in space and time, which is what is currently happening, smaller bees may have a greater ability to complete their life cycle than larger bees. Recent studies on different continents show that over the last century, large species are declining faster than smaller species (Bartomeus *et al.,* 2013; Scheper *et al.,* 2014).

This suggests that, under current conditions, greater mobility and/or amplitude in foraging is no longer compensating for greater demands on resources and that, for large bees, but not for small ones, there is a selective pressure for them to become smaller. This led to the hypothesis that during the last century, bee species with larger body sizes were becoming smaller over time, but bee species with smaller body sizes were not. This hypothesis was tested by measuring the body size of more than 4,500 specimens of 18 different species in entomological collections, all of which were collected in the Netherlands over a period of 149 years.

The Netherlands is particularly suitable for this type of research because it has extensive agricultural areas that serve as habitat for most bee species, and because, over the study period, these areas have gradually changed and become less and less suitable for bees (Kosior *et* al., 2007; Scheper *et al.,* 2014). During this period the bees have been collected by entomologists, making it possible to trace changes in their body size. Because bees generally show a visible sexual dimorphism in size, with females being significantly larger than males, we investigated whether this hypothetical relationship is the same in male and female bees.

Finally, because climate change affects both foraging opportunities (number of days that are warm and dry and thus more suitable for collecting pollen and nectar) and larval development (Radmachar & Strohm, 2009), we also assessed whether possible trends and changes in the bees' body size were influenced by temperature changes over time.

2.2. MATERIALS AND METHODS

2.2.1. Bee sampling and entomological collections

In this study, the intertegular distance (ITD) was used to assess the bees' body size. The ITD represents the distance between the two wing insertions and is an excellent estimate for assessing this characteristic (Cane, 1987). We therefore measured the DIT (mm) of bee specimens from the entomological collections of the Natural History Museum *(Naturalis Biodiversity Centre)* in Leiden, Holland (Figure 1). Bees are very well represented there, and for some species, body size measurements were possible using old specimens in the museum's extensive collection, some of which date back to 1866.

Figure 1. Entomological collection of the Natural History Museum (Naturalis Biodiversity Centre) in the province of Leiden, Netherlands.

The measurements were made on large and small bees, using a total of 18 species, divided into 7 different genera: *Andrena (A. barbilabris, A. bicolor, A. nitidá), Anthophora (A. plumipes, A. retusa), Bombus (B. pascuorum, B. pratorum, B. terrestris), Halictus (H. rubicundas, H. tumulorum), Lasioglossum (L. leucozonium, L. calceatum, L. villosulum), Megachile (M. centuncularis, M. leachella, M. marítima)* and *Osmia (O. caerulescens, O. rufa),* which are the best represented species in the collection. For the species of the genus *Bombus,* only queens and males were used. For the genera with two species sampled we selected one large and one small, and within the genera with three species sampled we selected one small, one intermediate and one large species (Figure 2).

A total of 4,510 specimens were measured, on average 131 females and 120 males per species (Table 1). In order to ensure equal distribution between the samples throughout the study period, the following intervals were defined: before 1900, 1900-1919, 1920-1939, 1940-1959, 1960-1979, 1980-1999, after 2000. The aim was to sample 20 individuals for each sex and species in each time interval.

Table 1: Species used in the research, number of individuals sampled for each species and sex, range in years and average size (mm) of bees collected in the Netherlands from 1866 onwards and kept in the entomological collections of the Natural History Museum (Naturalis Biodiversity Centre) in Leiden.

Species	Individuals sampled		Range in years		Average size (mm)		Size
	Female	Male	Female	Male	Female	Male	
Andrena barbilabris	140	140	1877-2010	1884-2007	2,28	1,71	Int
Andrena bicolour	140	110	1876-2007	1873 -2002	1,65	1,25	Small
Andrena nitida	135	134	1900-2009	1870-2009	2,96	2,19	Grd
Anthophora plumipes	115	94	1879-2009	1879-2003	4,55	4,34	Grd
Anthophora retusa	117	136	1875-2001	1874-2001	3,79	3,61	Small
Bombus pascuorum	151	164	1878-2006	1880-2000	4,84	3,89	Int
Bombus pratorum	136	165	1868-2005	1868-2005	4,73	3,51	Small
Bombus terrestris	126	111	1867-2007	1879-2003	5,48	3,85	Grd
Halictus rubicundus	142	113	1872-2002	1866-2001	2,05	1,72	Grd
Halictus tumulorum	119	110	1874-2012	1874-2001	1,10	1,00	Small
Lasioglossum calceatum	139	120	1866-2012	1866-2002	1,77	1,65	Grd
Lasioglossum leucozonium	128	113	1866-2008	1867-2013	1,92	1,56	Int
Lasioglossum villosulum	116	74	1868-2002	1878-2000	1,38	1,17	Small
Megachile centuncularis	127	116	1868-2004	1879-2000	2,96	2,69	Int
Megachile leachella	122	95	1869-2013	1869-2006	2,84	2,49	Small
Maritime Megachile	122	108	1871 -2002	1871 -2006	3,35	3,29	Grd
Osmia caerulescens	138	118	1871 -2002	1869-2002	1,98	1,67	Small
Osmia rufa	138	138	1869-2002	1866-2008	2,70	2,23	Grd

* Size: Int: intermediate size; Peq: small size; Gdr: large size.

To quantify the changes in temperature over the entire study period, we calculated, for each year, the average daily temperature during the flight periods of each species, for males and females separately. The Dutch national bee distribution database was used (Reemer *et al.,* 2012) and based on the recordings from this database, the flight period of each species was calculated (Scheper *et al.,* 2014). The average daily temperature during the flight period of the bees was calculated using the extensive collection of meteorological data (1901 - 2012) obtained from the De Bilt weather station *(Royal Netherlands Meteorological Institute)* located in central Holland. There is no meteorological data for the period before 1901.

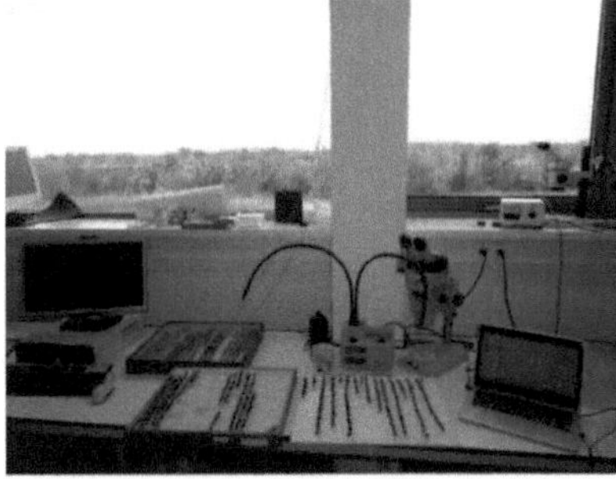

Figure 2. Intertegular distance (ITD) used to assess the bees' body size and the data collection laboratory at the Alterra Institute at Wageningen University in the Netherlands.

2.2.2. Statistical analyses

This study used Linear Mixed Modelling (LMMs) to examine whether body size decreased faster in large bees than in small bees. 'DIT', our variable used for body size, was transformed into a logarithm (LoglO) to reduce error heterogeneity. In addition, species was included as a random factor, which represents potential dependence of trends and characteristics between such closely related bee species (Bartomeus *et al.*, 2013). Phylogenetic regression was not used, as data on phylogenetic analyses are still controversial (Westoby *et al.*, 1995; Hedtke *et al.*, 2013).

An initial analysis was carried out including the fixed factors 'Initial Size', 'Year', 'Sex', as well as all double and triple interactions. The Initial Size was defined as the average DIT (Intertegular Distance) of bees collected before 1900 (1866- 1899). Although we usually have the exact day of capture information for most specimens, only the year of capture was used in the analyses. This initial analysis showed that female and male bees show the same trend in body size over the years, with evidence that the relationship between measured Body Size and Initial Size differed between the sexes (interaction Initial Size*Gender, E^is.g = 267.77, P<0.001). In view of this difference, it was decided to carry out separate analyses for males and females.

These LMMs included Initial Size, Year and their interactions as fixed terms. In these analyses, the statistical significance of the negative interactions between Initial Size and Year was interpreted as support for the hypotheses raised. Because temperature data was only available from

1901 onwards, a second analysis was carried out which included the average temperature during the flight period of the year prior to the capture of each individual. The reason for using the average temperature during the flight period of the previous year is that temperate bee species hibernate as pre-pupae or adults and emerge from their brood cells in the year following oviposition.

Again, separate analyses were made for males and females and LMMs were used to relate bee body size to Initial Size, Year, Temperature in the year prior to measurements and all interactions. Here, the statistical significance of the interaction between Body Size and Temperature was used as an indicator that temperature affects trends in bee body size (DIT).

All the models were examined using tools from the GenStat statistical programme (Payne *et al.*, 2002).

2.2.2.1. Analysing the decline in body size of bees of the genus *Bombus*

To analyse the decline in body size of the species of the genus *Bombus* separately, an Analysis of Variance (ANOVA) was carried out using the statistical programme PAST version 2.02, with the data compared *a posteriori* using the Tukey-Kramer multiple comparison test.

This analysis also used data on the body size of workers. Therefore, to measure the body size of bees of the genus *Bombus,* our main object in this thesis, we collected data on the Intertegular Distance (ITD) of 1,293 specimens, including workers, females (new queens) and males, and divided into three different species *Bombus pascuorum, Bombus pratorum* and *Bombus terrestris,* between the time period of 1867 and 2007.

2.3. RESULTS AND DISCUSSION

There is a strong relationship between the body size of bees and the process of natural and agricultural pollination, and many studies suggest advantages for large bees in foraging efficiency (Klostermeyer *et al.,* 1973; Pyke, 1978; Torchio & Tepedino, 1980). However, according to the results, since 1866 the 18 bee species studied have shown a decline in body size for both sexes (Figure 3), this average size reduction was 4.9 per cent.

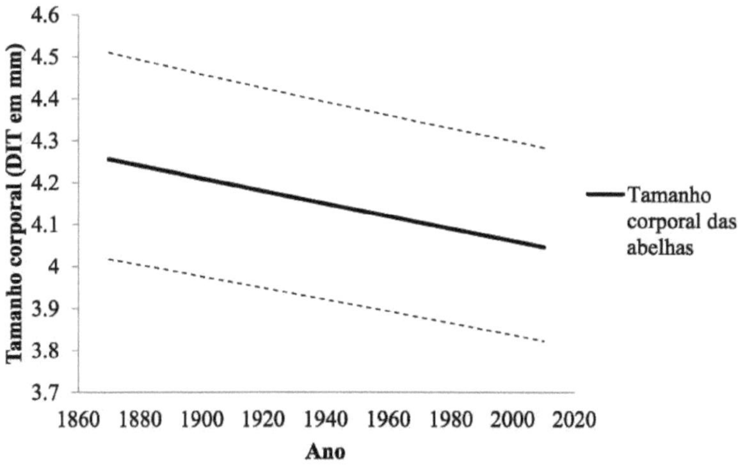

Figure 3: Predictions from the first analysis (transformed into original values) of the body size of all 18 bee species (Intertegular Distance) collected in the Netherlands between 1866 and 2013.

Table 2. Results of the Linear Mixed Model (LMMs) testing the relationship between Initial Size, Year and Sex and all interactions with Body Size (DIT) of the bee species used in the research. Species and gender were included as a random term in the model, n.g.l. numerator degrees of freedom, d.g.l denominator degrees of freedom.

Fixed Terms	n.d.f	d.d.f.	Statistician	P
Initial size	1	3853,5	4943,46	<0,001
Year	1	4485,1	57,54	<0,001
Sex	1	4287,1	456,23	<0,001
Initial Size*Gender	1	4418,9	267,77	<0,001
Initial size* Year	1	4485,2	5,28	0,022
Year*Gender	1	4484,9	0,00	0,987
Initial Size*Year*Gender	1	4485,3	0,07	0,78

Females were on average 19.7% (s.e. = 2.85%, n = 18) larger than males ($F_{1; 4287,1}$ = 456.23, P<0.001) and the sex-specific analysis showed that the body size of females declined significantly over approximately 150 years of the study period ($F^\wedge{}_{2332.2}$ = 40.30, P<0.001, Table 3). However, large females had a more pronounced decline in body size than small females (Figure 4). Between 1866 and 2012, on average for all species, the predicted decline in "DIT" was 2.3 per cent and 7.8 per cent for females with an Initial Size of 1.7 and 6.0 mm respectively.

Males similarly showed a total decline of 2.7 per cent in their body size since 1866 ($F_{1,2138.7}$= 12.20, P<0.001), however, unlike females, large and small males had no significant difference and showed a similar trend in body size decline (Initial Size*Year interaction: $F_{1; 2339.3}$ = 0.00, P = 0.986, Figure 4).

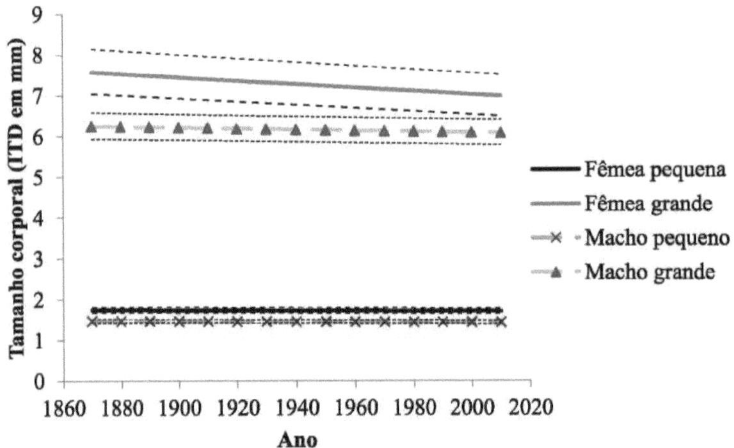

Figure 4 - Predictions from the second analysis (transformed into original values) of body size (Intertegular Distance), of large and small females and large and small males separately, in the Netherlands during the period between 1866 and 2013.

Table 3. Results of the Linear Mixed Model (LMMs) testing the relationship between Initial Size, Year and their interactions with the body size of females and males separately. Species and gender were included as a random term in the model, n.g.l. numerator degrees of freedom, d.g.l denominator degrees of freedom.

Fixed terms	n.d.f.	d.d.f.	Statistician	P
Females				
Initial size	1	6,2	233,75	<0,001
Year	1	2332,2	40,30	<0,001
Initial Size*Year	1	2332,2	7,51	0,006
Males				
Initial size	1	4,2	479,51	<0,001
Year	1	2138,7	12,20	<0,001
Initial Size*Year	1	2139,3	0,00	0,986

Changes in the body size of bees, especially in larger species, as the results show, can interfere with the relationship between bees and flowers. Larger species of bees that are large and have a large body size, especially females, which are larger and collect more resources (nectar and pollen), have a higher food requirement than smaller species, but over the last few decades man has been interfering a lot with the environment, and areas that once served as natural habitat for bees are now being destroyed. Changes in land use, extensive areas for livestock farming and agricultural intensification are considered the main causes of the loss of wild bee diversity. These environmental changes have caused and are still causing a drastic reduction in available floral resources, interfering with the breadth of the diet, pollen collection and floral preference of bee species and causing changes in population trends and traits such as body size, for example (Foley *et al.*, 2005; Green *et al.*, 2005; FAO, 2006; Potts *et al.*, 2010; Winfree *et al.*, 2011).

Many studies attribute high foraging efficiency to large females. These females can collect a greater amount of pollen and nectar, visit more flowers per unit of time and can fly long distances in search of food (Pyke, 1978; Gathmann *et al.*, 1994; Strohm & Linsenmair, 1997; Tomkins *et al.*, 2001; Pereboom & Biesmeijer, 2003).

These bees are directly related to plant reproduction and agricultural production, but because of all these environmental changes the larger bees are probably suffering selective pressure to become smaller.

According to our predictions, larger females are becoming smaller, more so than smaller females, reinforcing a recent study in the Netherlands, which showed that large bee species are suffering a strong decline, greater than smaller bees, during the last century (Scheper & Kleijn, 2014), the same pattern was found in the USA (Bartomeus *et al.*, 2013).

Females require larger food supplies than males (Maddocks & Paulus, 1987; Strohm, 2000; Strohm *et al.*, 2002), and because of their high pollen requirement, larger species are more susceptible to changes in land use and climate change (Bartomeus, *et al.*, 2013; Larsen *et al.*, 2005; Muller *et al.*, 2006).

In solitary bees, body size can be influenced by the quantity of larval provisions, the quality of the provisions and temperature during development (Klostermeyer *et al.*, 1973; Bosch & Vicens, 2002), but temperature during development is probably the most important factor in determining body size (Radmacher & Strohm, 2009; Atkinson, 1994).

Due to climate change, the average temperature during the period of the year when bees are active and flying (spring and summer) has increased significantly over the past century in the Netherlands (Scheper *et al.*, 2014) (Figure 5). There are two possible explanations for the effect of temperature on bees. Firstly, it is an indicator for the number of warm, sunny days in temperate regions, which have a major influence on the bees' foraging and food supplies for their brood or colonies. Better supplies generally result in larger broods (Klostermeyer *et al.*, 1973, Bosch & Vicens, 2002). A second explanation is that high temperatures during larval development have been reported to negatively affect the body size of bees (Radmacher & Strohm 2009).

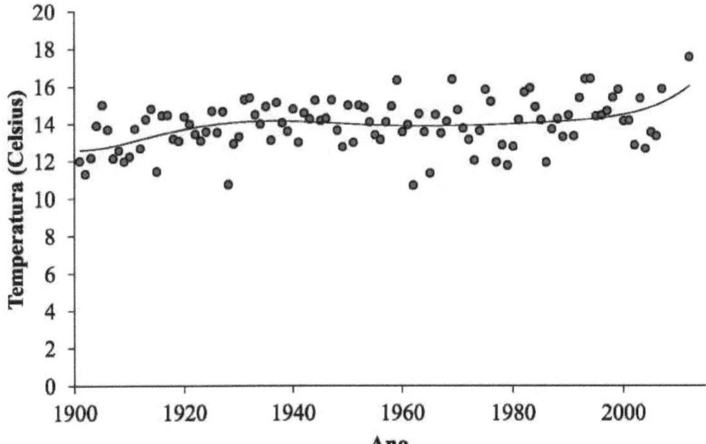

Figure 5 - Average temperature (Celsius) during the flight period of the 18 species of small and large bees used in the research and collected in the Netherlands between 1866 and 2013.

According to Atkinson (1994) and Radmacher & Strohm (2009) low temperatures during development result in larger adult bees and the weight of the cocoon containing the brood decreases significantly with the increase in temperature. In this research, the average temperature during the bees' flight period increased over time (Figure 5). However, incorporating the average temperature during the flight period of the year before each measurement did not alter the results (Table 4). Temperature had no measurable relationship with the body size of the bees studied.

In females, with the incorporation of Temperature into the analyses, the difference in response between small and large bees remained as indicated by the high significance of the Year*Initial Size interaction (Table 3). In males, this interaction remained non-significant, indicating a similar trend in the body size of small and large bees.

Table 4: Results of the Linear Mixed Model (LMMs) testing the relationship between Initial Size, Year and Temperature and all interactions with bee body size. Species and gender were included as a random term in the model, n.g.l. numerator degrees of freedom, d.g.l denominator degrees of freedom.

Fixed terms	n.d.f	d.d.f.	Statistician	P
Females				
Initial size	1	6,0	230,46	<0,001
Year	1	2050,4	27,05	<0,001
Temperature	1	1885,5	0,6	0,437
Year*Temperature	1	2045,4	2,1	0,147
Initial Size *Year	1	2051,1	11,82	<0,001
Initial Size*Temperature	1	2023,3	0,7	0,404
Initial Size*Year*Temperature 1		2044,2	0,72	0,396
Males				
Initial size	1	3,7	448,06	<0,001
Year	1	1843,4	5,85	0,016
Temperature	1	640,1	0,31	0,58
Year*Temperature	1	1899,2	5,7	0,017
Initial Size*Year	1	1874,5	1,61	0,204
Initial Size*Temperature	1	935,1	1,9	0,169
Initial Size*Year*Temperature 1		1899,6	2,52	0,113

2.3.1. Decline in body size of bees of the genus *Bombus*

Of the three species of *Bombus* analysed, *B. pratorum* did not show significant reductions in body size, but the other two species, *B. pascuorum* and *B.*

terrestris showed significant reductions in body size. Bees of the species *B. pascuorum* had an extremely significant body decline over time (p<0.0001) for all castes (Figure 6).

Figure 6 - Original body size values (DIT in mm) of workers, females (new queens) and males of the species *B. pascuorum* collected in the Netherlands between 1867 and 2007.

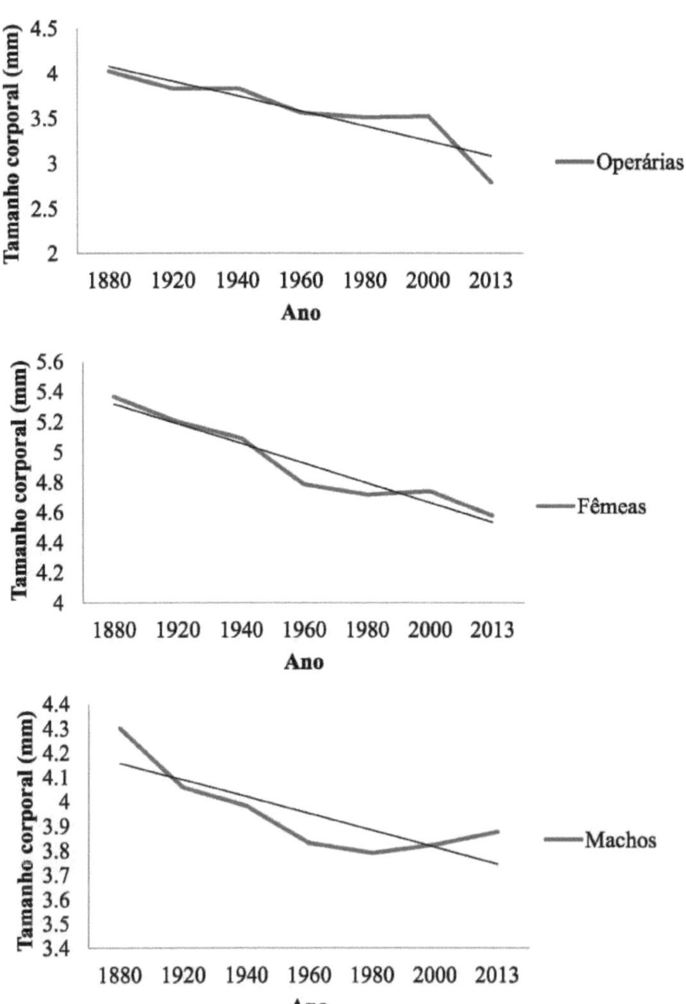

The reduction in body size of the *B. pratorum* species was not significant, but the same was not true of the *B. terrestris* species, which showed a decline in body size similar to that of the *B. pascuorum* species (Figure 7). Bees of this genus are extremely important, both in maintaining biodiversity and in the natural pollination that occurs in agricultural areas (Velthuis & Doom, 2006).

They are considered to be very effective pollinators, precisely because of their large body size. They are also specific pollinators of various crops such as tomatoes *(Solanum lycopersicum)* and peppers *(Capscum annuum),* for example, which have poricidal anthers and benefit from the *buzz* pollination behaviour displayed by this type of bee. All the environmental changes caused by man are having severe negative effects on bee populations and are interfering with important characteristics and tendencies for pollination, such as body size.

Figure 7. Original body size values (DIT in mm) of workers, females (new queens) and males of the species *B. terrestris* collected in the Netherlands between 1867 and 2007.

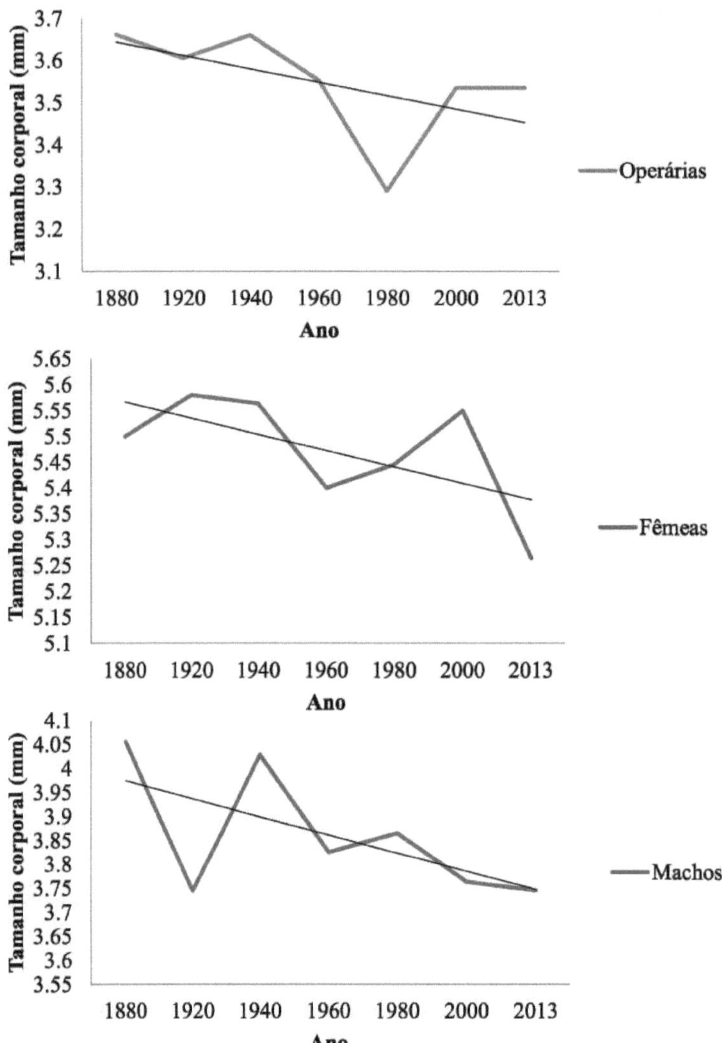

Many studies point to the genus *Bombus* as one of the most affected today (Goulson, 2003; Biesmeijer *et al.*, 2006; Kosior *et al.*, 2007) and changes in certain specific characteristics in these large bees, as shown by the results of this research, serve as an indication of the decline of these bees and point to the urgent need to seek solutions to these problems that contribute to the conservation of bee species, as well as pollinator species as a whole.

2.4. CONCLUSIONS

According to the results, it can be concluded that during the approximately 150 years of the study period the 18 species investigated showed a decline in body size, and changes in this characteristic can have severe implications for the flower bee relationship.

The reduction in body size found can interfere with natural pollination that occurs in agricultural areas, as well as plant biodiversity, since the relationship between bees and flowers is highly dependent on and influenced by the body size of the pollinator.

The sharp decline in the body size of large females can be passed on to future generations and serves as an indication of the selective pressure suffered by larger bees.

The genus *Bombus,* due to the large size of its species, is being particularly affected by environmental changes, increasing the need and importance of more research into these bees in order to generate new information on each individual species and, perhaps, develop a rational breeding model for these bees that will allow them to be managed, used for pollination and the introduction of colonies in places previously populated by them.

2.5. BIBLIOGRAPHICAL REFERENCES

AGRAWAL, ANURAG A.; HASTINGS, AMY P.; JOHNSON, MARC T. J.; MARON, J. L. & JUHA-PEKKA SALMINEN J. P. Insect Herbivores Drive Real-Time Ecological and Evolutionary Change in Plant Populations. **Science,** v. 338, n. 6103, p. 113-116, 2012.

ATKINSON, D. Temperature and organism size - a biological law for ectotherms? **Advances in Ecological Research,** v. 25, p. 1-58,1994.

BARTOMEUS, L; ASCHER, J.S.; GIBBS, J.; DANFORTH, B.; WAGNER, D.; HEDKE, S.H. & WINFREE, R. Historical changes in northeast US bee pollinators related to shared ecological traits. PNAS, v. 110, n. 12, p. 4656-4660,2013.

BIESMEUER, J.C.; ROBERTS, S.P.M.; REEMER, M.; OHLEMÚLLER, R.; EDWARDS, M.; PEETERS, T.; SCHAFFERS, A.P.; POTTS, S.G.; KLEUKERS, R.; THOMAS, C.D.; SETTELE, J. & KUNIN, W.E. Parallel declines in pollinators and insect-pollinated plants in

Britain and the Netherlands. **Science,** v. 313, p. 251-353, 2006.

BOMMARCO, R.; LUNDIN, O.; SMITH, H. G. & RUNDLÔF, M. Drastic historical shifts in bumble-bee community composition in Sweden. **Proc. R. Soc. B,** v. 279, p. 309-315, 2012.

BOSCH, J. & VICENS, N. Body size as an estimator of production costs in a solitary bee. **Ecological Entomology,** v. 27, p. 129-137, 2002.

BRADSHAW, W. E. & HOLZAPFEL, C. M. Evolutionary Response to Rapid Climate Change. **Science** v. 312, p. 1477-1478, 2006.

CANE, J. H. Estimation of bee size using intertegular span (Apoidea). **Journal of the Kansas Entomological Society,** v. 60, p. 145-147, 1987.

DEGUINES, N.; JONO, C.; BAUDE, M.; HENRY, M.; JULLIARD, R. & FONTAINE, C. Large-scale trade-off between agricultural intensification and crop pollination services. **Frontiers in Ecology and the Environment,** v. 12, n. 4, p. 212- 217, 2014.

FAO. **Livestock's Log Shadow.** 416p., 2006.

FOLEY J. A., *et al.* Global consequences of land use. **Science,** v. 309, n. 5734, p. 570- 574, 2005.

FREITAS, B.M.; IMPERATRIZ-FONSECA, V.L.; MEDINA, L.M.; KLEINERT, A.M.P.; GALETTO, L.; NATES-PARRA, G. & QUEZADA-EUÁN, J.J.G. Diversity, threats and conservation of native bees in the Neotropics. **Apidologie,** v. 40, p. 332-346, 2009.

GATHMANN, A.; GREILER, H. J. & TSCHARNTKE, T. Trap nesting bees and wasps colonising set aside fields: Succession and body size, management by cutting and sowing. **Oecologia,** v. 98, p. 8-14,1994.

GOULSON, D. Bumblebees: **Behaviour and Ecology.** Oxford, UK: Oxford Univ. Press, 2003.

GREEN, R.E.; CORNELL, S.J.; SCHARLEMANN, J.P.W. & BALMFORD, A. Farming and the fate of wild nature. **Science,** v. 307, n. 5709, p. 550-555, 2005.

GREENLEAF, S.S.; WILLIAMS, N.M.; WINFREE, R. & KREMEN, C. Bee foraging ranges and their relationship to body size. **Oecologia,** v. 153, n. 3, p. 589-596, 2007.

HEDTKE, S.M.; PATINY, S.& DANFORTH, B.N. The bee tree of life: a supermatrix approach to apoid phylogeny and biogeography. **BMC Evolutionary Biology,** v. 13, p. 138, 2013.

HUGHES, L. Biological consequences of global warming: is the signal already apparent? **Trends in Ecology and Evolution,** v. 15, p. 56,2000.

KEVAN, P.G. & PHILLIPS, T.P. The economic impacts of pollinator declines: an approach to assessing the consequences. **Conservation Ecology,** v. 5, p. 1: 8, 2001.

KLEIJN, D. & RAEMAKERS, I. A retrospective analysis of pollen host plant use by stable and declining bumblebee species. **Ecology,** v. 89, p. 1811-1823,2008.

KLEIN, A.M.; VAISSIERE, B.E.; CANE, J.H.; STEFFAN-DEWENTER, L; CUNNINGHAM, S.A.; KREMEN, C. & TSCHARNTKE, T. Importance of pollinators in changing landscapes for world crops. **Proceedings of the Royal Society B: Biological Sciences,** v. 274, p. 303-313, 2007.

KLOSTERMEYER, E.C.; MECH, S.J. & RASMUSSEN, W.B. Sex and weight of Megachile rotundata (Hymenoptera: Megachilidae) progeny associated with provision weights. **Journal of the Kansas Entomological Society,** v. 46, p. 537-548, 1973.

KREMEN, C; WILLIAMS, N.M. & THORP, R.W. Crop pollination from native bees at risk from agricultural intensification. **Proceedings of the National Academy of Sciences of the United States of America,** v. 99, p. 16812-16816, 2002.

LARSEN, T.H.; WILLIAMS, N.M. & KREMEN, C. Extinction order and altered community structure rapidly disrupt ecosystem functioning. **Ecology Letters,** v. 8, n. 5, p. 538-547, 2005.

MADDOCKS, R. & PAULUS, H.F. Quantitative Aspekte der Brutbiologie von Osmia rufa L. und Osmia comuta Latr. (Hymenoptera: Megachilidae): Eine vergleichende Untersuchung zu Mechanismen der Konkurrenzminderung zweier nahverwandter Bienenarten. **Zool. Jahrb. Abt. Syst. Ôkol. Geogr. Tiere,** v. 114, p. 15 44,1987.

MULLER, A.; DIENER, S.; SCHNYDER, S.; STUTZ, K.; SEDIVY, C. & DORN, S. Quantitative pollen requirements of solitary bees: implications for bee conservation and the evolution of bee-flower relationships. **Biological Conservation,** v. 130, n. 4, p. 604-615,2006.

OLLERTON, J; WINFREE, R & TARRANT, S. How many flowering plants are pollinated by animals? **Oikos,** v. 120, n. 3, p. 321-326, 2011.

OLROYD, B.P. What's killing American honey bee? **PLoSBiology,** v. 5, n.6, p.168, 2007.

PAYNE, R.W.; BAIRD, D.B.; CHERRY, M.; GILMOUR, A.R. & HARDING, S.A. **Genstat for Windows,** 6th edn VSN International, Oxford, UK, 2002.

PEREBOOM, J. J. M. & BIESMEIJER, J. C. Thermal constraints for stingless bee foragers: the importance of body size and colouration. **Oecologia,** v. 137, p. 42-50, 2003.

POTTS, S. G.; BIESMEIJER, J. C.; KREMEN, C.; NEUMANN, P.; SCHWEIGER, O. & KUNIN, W. E. Global pollinator declines: trends, impacts and drivers. **Trends Ecol** Evol, v. 25, n. 6, p. 345-353, 2010.

PYKE, G.H. Optimal body size in bumblebees. **Oecologia,** v. 34, p. 255-266,1978.

RADMACHER, S. & STROHM, E. Factors affecting offspring body size in the solitary bee *Osmia bicornis* (Hymenoptera, Megachilidae). **Apidologie,** v. 41, p. 169-177, 2009.

REEMER, M.; KLEIJN, D. & RAEMAKERS, I.P. **Changes in the Dutch bee fauna** (in Dutch with English summary). The Dutch Bees (Hymenoptera: Apidae s.L), eds Peeters, T.M.J., et al. (Invertebrate Survey - The Netherlands, Leiden), pp. 103-107, 2012.

ROYAL NETHERLANDS METEOROLOGICAL INSTITUTE (KNMI). Klimatologie: Daggegevens van het weer in Nederland (http://www.knmi.nl/klimatologie/daggegevens/download.html).

SCHEPER, J. & KLEIJN, D. **Mitigating pollinator loss in Europe: what strategies are most effective?** PhD Thesis. Natural Conservation and Plant Study Group, Wageningen University, 2014.

SCHEPER, J.; REEMER, M.; VAN KATS, R.; OZINGA, W.A.; VAN DER LINDEN, G.T.J.; SCHAMINÉE, J.H.J.; SIEPEL, H. & KLEIJN, D. Museum specimens reveal loss of pollen host plants as key factor driving wild bee decline in the Netherlands. **Proceedings of the National Academy of Science of the United States of America** v. 111, p. 17552-17557, 2014.

STROHM, E. & LINSENMAIR, K. E. Low resource availability causes extremely male-biased investment ratios in the European beewolf, *Philanthus triangulum* F. (Hymenoptera, Sphecidae). **Proceedings of the Royal Society of London B,** v. 264, p. 423-429, 1997.

STROHM, E. Factors affecting body size and fat content in a digger wasp. **Oecologia,** v. 123, p. 184-191,2000.

STROHM, E.; DANIELS, H.; WARMERS, C. & STOLL, C. Nest provisioning and a possible cost of reproduction in the megachilid bee *Osmia rufa* studied by a new observation method. **Ethology Ecology & Evolution,** v. 14, p. 255-268, 2002.

STROHM, E. & LINSENMAIR, K. E. Low resource availability causes extremely male-biased investment ratios in the European beewolf, Philanthus triangulum F. (Hymenoptera, Sphecidae). **Proceedings of the Royal Society of London B,** v. 264, p. 423-429, 1997.

TOMKINS, J. L.; SIMMONS, L. W. & ALCOCK, J. Brood-provisioning strategies in Dawson's burrowing bee, *Amegilla dawsoni* (Hymenoptera: Anthophorini). **Behavioral Ecology and Sociobiology,** v. 50, p. 81-89, 2001.

TORCHIO, P. F. & TEPEDINO, V. J. Sex ratio, body size and seasonality in a solitary bee, *Osmia lignaria* propinqua Cresson (Hymenoptera: Megachilidae). **Evolution,** v. 34, p. 993-1003,1980.

VELTHUIS, H.H.W. & DOORN, A. van. A century of advances in bumblebee domestication and the economic and environmental aspects of its commercialisation for pollination. **Apidologie,** v. 37, p. 421-451, 2006.

WESTOBY, M; LEISHMAN, M. R & LORD, J. M. On misinterpreting the 'phylogenetic correction'. **Journal Ecology,** v. 83, n. 3, p. 531-534,1995.

WINFREE, R.; BARTOMEUS, L; CARIVEAU, D. P. Native pollinators in anthropogenic habitats. **Annual Review of Ecology, Evolution, and Systematics,** v. 42, p. 1-22, 2011.

CHAPTER 3

Nesting biology and rational breeding of *Bombus (Thoracobombus) brevivillus* in north-eastern Brazil Nesting biology and rational breeding of *Bombus (Thoracobombus) brevivillus* in north-eastern Brazil

Abstract - This research describes the nesting behaviour and the colony of *Bombus (Thoracobombus) brevivillus,* and provides information on the rational breeding of this Neotropical species, which only occurs south of the equator. We found and observed seven colonies of *B. brevivillus* in the state of Ceará, north-east Brazil. Four of the seven colonies were captured and transferred to wooden boxes that allowed for their rational rearing, with the aim of obtaining information on the pollen types used, the foraging pattern, resource collection, the influence of climatic variations on the external activities of the workers and on the development and establishment of the colonies in rational boxes. Observations have shown that *B. brevivillus* seems to be opportunistic when it comes to nesting sites and protecting the young, and does not invest much in building the nest and protecting the offspring. It can be seen that *Bombus brevivillus* invests a lot of effort in producing new young and developing colonies, before producing reproductive individuals. The colonies therefore produce a large number of workers, a characteristic that can be useful for commercial pollination. Among the plants visited by the workers, the species *Chamaecrista duckeana* (Fabaceae Caesalpinioideae) and *Tabebuia impetiginosa* (Bignoniaceae) were the most representative in the bees' protein diet under natural conditions and under rational breeding, respectively. Both the flow of bees and the collection of pollen and nectar/water were higher in the early hours of the day. Climatic variations influenced the external activities of the workers; in the hottest hours they reduced their foraging and pollen collection activities and began the process of thermoregulation. At the end of the biological cycle, one of the colonies was reactivated by a queen from the same nest, showing that, unlike temperate species, *B. brevivillus* can have perennial colonies.

Keywords: Colony size, nest characteristics, foraging pattern, resource collection.

3.1. INTRODUCTION

Bombus is a primitive genus of eusocial bees, with approximately 250 species described worldwide. In South America there are only 22 known species divided into three subgenera (Hines *et al.,* 2007; Williams *et al.,* 2008). Tropical species, mainly Neotropical, are less abundant (Moure & Melo, 2012). In Brazil, the genus is represented by only six species, all belonging to the subgenus *Thoracobombus'. B. bellicosus, B. brasiliensis, B. brevivillus, B. morio, B. atratus* and *B.*

transversalis (Moure & Melo, 2012). Of these, the species *B. brevivillus* has very little information in the literature.

The *Bombus brevivillus* species is widely distributed in Brazil, Guyana and Venezuela (Moure & Melo, 2012), but is little studied. Like many other Neotropical species, this species is becoming scarce in some areas due to accelerated deforestation and intensification of agriculture (Kleijn & Raemakers, 2008; Freitas *et al.,* 2009), but also because farmers constantly destroy nests to prevent attacks by workers (Garófalo, 2005). As a result, wild nests are difficult to find and to gain access to inside colonies, so the biology of *B. brevivillus* is very poorly understood.

Bees of this genus are widely used for pollination in protected cultivation and there is a great demand for them to be bred on a large scale for this purpose. Producers all over the world use *Bombus* bees for pollination in protected cultivation (Velthuis & Doom, 2006). Increases in production reported with their use in tomato greenhouses, for example, can exceed 28% and the fruit obtained is all of superior quality (Sande, 1990; Fiume & Parisi, 1994).

The use of protected cultivation in Brazil has expanded considerably, especially since the early 1990s (Grande *et al.,* 2003). According to data obtained in 2011 by the Brazilian Committee for the Development and Application of Plastics in Agriculture (Coblapa), it is estimated that production in protected environments in Brazil occupies around 26,000 hectares. With this form of cultivation on the rise, the use of these bees would be very important for improving the productivity of our crops, as is the case in much of Europe. However, the introduction of exotic species can have severe negative effects and is prohibited by Brazilian law (Saraiva *et al.,* 2012).

In Brazil, studies on *Bombus* bees are very scarce, and there is not much information on the *B. brevivillus* species. It is therefore necessary to study the species that exist here in order to gain a better understanding of their characteristics, and to develop management techniques that allow them to be bred rationally, as we will certainly find efficient pollinators for many of our crops among them, such as the *Bombus brevivillus* bee, for example.

Therefore, the aim of this study is to describe the nesting behaviour and colony of *Bombus (Thoracobombus) brevivillus,* and to investigate the habits of this species, such as the types of pollen used by this bee, to evaluate its foraging pattern, resource collection, its biological cycle and the influence of climatic variations on the external activities of workers kept in rational breeding conditions.

3.2. MATERIALS AND METHODS

3.2.1. Areas of study

The study was carried out in two different locations, firstly in the municipality of Itatira, state of Ceará, Brazil (04°37'32.83"S 39° 34'12.11"W) where the rainy season runs from January to the end of April, with a little or no rain in the other months. The average annual rainfall is 807.8 mm and the average annual temperature is 25°C (IPECE, 2012). The second study area was in an urban area, in the Bee Sector of the Federal University of Ceará, located in the municipality of Fortaleza (03°44'45.28"S 38° 34'30.13"O). This region has a warm, tropical climate, similar to that which occurs along the entire coast of Brazil. The average annual temperature is 26°C, with the highest temperatures occurring between December and January (IPECE, 2012).

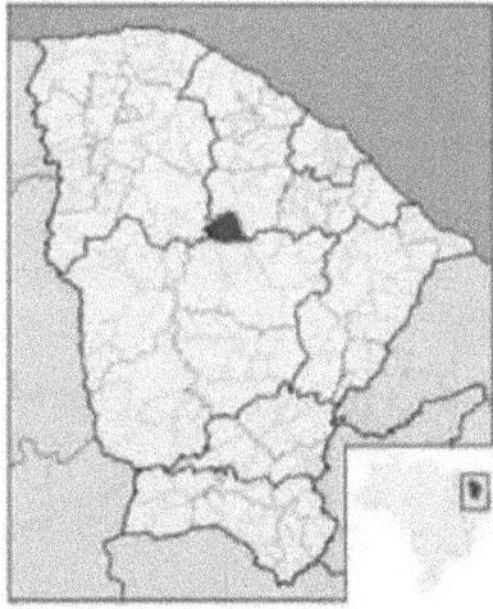

Figure 8. Geographical location of the study areas, showing the state of Ceará in north-eastern Brazil and the location of the municipality of Itatira.

3.2.2. Nesting behaviour of *Bombus brevivillus*

The observations were carried out in the municipality of Itatira (04°37'32.83"S 39° 34' 12.11' 'O). Several active searches were carried out throughout the municipality and seven colonies of *B. brevivillus* were found in and around a banana plantation *(Musa paradisíaca)*. At an altitude ranging from 323 to 849 metres. Here, observations were made regarding the location of the nest, the type of protection for the cocoons, the number of entrances and the global position (GPS) of each nest found.

3.1.1. Capturing colonies of *Bombus brevivillus*

Four of the seven nests found were opened, using reinforced beekeeping clothing to minimise stings. The leaf cover was carefully removed by hand or the nest was excavated into a cavity to gain access to the brood envelope and cocoons. All four colonies had large numbers of workers and had already started laying reproductive individuals. These colonies were captured and transferred to rational wooden brood boxes (50 cm long x 30 cm wide x 40 cm high) with a glass slide under the wooden cover to allow observations. All the individuals, including the workers returning from foraging, were captured and transferred to the brood boxes which were left open in the same original location as the nest.

Only in the evening, when there were no more bees flying or in sight, were the boxes closed for good and transferred by car the same night to the Bee Stor at the Department of Animal Science at the Federal University of Ceará in Fortaleza, Brazil (03°44'45.28"S 38° 34'30.13"O). We didn't check the nest sites the next morning to see if any bees had been left behind. Once at the University, the rational boxes were placed on the ground, protected by mutre plants *(Aloysia virgatal),* and left open to allow the bees to forage naturally from the following morning onwards.

Figure 9. (a) Model of the box used in the rational breeding of the native species *Bombus brevivillus-,* (b) use of beekeeping equipment and clothing to capture the colonies; (c) detail of the cocoons after removing the leaves at the time of capture; (d) transfer of the entire colony, including the queen, to the rational box; (e) box left in the same place as the nest until the evening, allowing all the workers to return; (f) already installed in the Bee Sector of the Federal University of Ceará; (g) artificial feeders and a glass slide over the box, which allowed observations and monitoring of the colonies;

46

(h) one of the colonies seen from above, at the time of the nocturnal revisits.

3.1.2. Description of the *Bombus brevivillus* colony

Information was collected on the location of the nest, the type of cocoon protection, the number of entrances and the geographical coordinates (GPS) of each nest found. The external activities of each nest were observed for a whole day before carrying out any intervention on the nests. Photographs and notes were taken before removing the envelope and exposing all the young and pots. The number of honey/pollen pots, the number of adult bees and the number of brood cells were counted and recorded (no distinction was made between cells containing larvae or pupae). The ratio of brood to adult bees in each nest was also calculated.

3.1.3. Foraging patterns, resource collection and the influence of climatic variations on the external activities of workers kept in rational breeding conditions.

Two of the four colonies captured were used to determine the foraging activities of the worker bees. Weekly observations were made at two-day intervals, in the first ten minutes of each hour, starting at 5am and ending at 6pm, in order to indicate the times of greatest bee flow (preferred time for collecting floral resources), the amount of bees collecting pollen and nectar or water and the influence of climatic variations on the bees' external activities. This was done by counting the number of bees going out to forage, arriving from the field, whether or not they were carrying pollen loads in their corbiculae and the number of bees ventilating the entrance to the rational box.

The identification of pollen in the corbicula (extension of the tibia of the third pair of legs present in workers, used to transport pollen to the box) of workers returning from the field was obtained by direct observation, taking into account the pollen's own characteristics (granular and generally coloured), while nectar or water loads were considered when the bees returned from the field without any material in the corbicula, according to the methodology described by Carvalho-Zilze *et al.* (2007).

To assess the influence of climatic variations and meteorological factors on the flight activity of the workers, temperature and relative humidity were recorded at one-hour intervals using an automatic weather station at the Federal University of Ceará, located approximately 500 metres from where the colonies were set up.

3.1.4. Types of pollen used by the native bee *Bombus brevivillus*

The feeding preferences of the *Bombus brevivillus* species were assessed during the months of September, October and November, and in two different ways: a) under natural conditions - pollen samples were collected directly from the storage jars of colonies captured in the municipality of Itatira, where, through palynological analyses, the pollen types found were compared with samples of the plant species that were in bloom at the time. A total of 4,096 pollen grains were counted on 13 microscope slides, a count of 300 pollen grains/slide. b) in rational breeding conditions - at seven-day intervals, from 6am to 1am, in the first ten minutes of each hour, all the workers with pollen loads visible on the corbiculae, who returned from the field to the boxes installed in the UFC's Bee Sector, were collected and the pollen grains were taken directly from the corbiculae. A total of 2088 pollen grains were counted on seven slides, an average of 300 pollen grains per slide.

The samples were freshly mounted on optical microscopy slides using the Wodehouse method (1935) in glycerine gelatine. The botanical origin of the pollen collected by the workers was characterised by microscopic analysis, where the pollen grains were identified by comparing them with the pollen types described in the literature (Moore & Webb, 1978), with the reference laminarium of the plant species collected in the municipality of Itatira, Ceará, which were in bloom during the research period and with the reference laminarium of the Bee Sector of the UFC.

The samples were analysed using two different methods: a) Qualitative method - the pollen types present were determined by comparison with the reference laminarium and the descriptions obtained from specialised literature (Barth, 1990) and b) Quantitative method - by counting 300 pollen grains/repeat/sample consecutively, where a total of 6184 pollen grains were counted to determine the classes of occurrence, which according to Louveaux *et al.* (1978) are:
PD (Dominant Pollen) - > 45% of the total number of grains; PA (Accessory Pollen) - 16 to 45% of the total number of pollen grains counted; PII (Important Isolated Pollen) - 3 to 15% and PIO (Occasional Isolated Pollen) - > 3%.

3.1.5. Development and establishment of colonies and activity of *Bombus brevivillus* queens in rational breeding conditions

The four colonies captured in the municipality of Itatira-CE and installed in the Bee Sector of the Federal University of Ceará were monitored for approximately four months. During the weekly reviews, information was obtained on the development and establishment of these colonies in captive conditions, the number of active workers throughout the period and the activity of the *B. brevivillus* queens.

3.1.6. Statistical analyses

INSTAT 3 was used for the statistical analyses, and the data relating to the foraging pattern (parametric data) was analysed by analysis of variance and compared *a posteriori* using the Tukey test. In order to statistically relate foraging activities to climatic conditions, linear correlations (r) were carried out using the statistical programme PAST version 2.02.

3.3. RESULTS AND DISCUSSION

3.3.1. Nesting behaviour of *Bombus brevivillus*

Observations of the nests of *B. brevivillus* showed that this species nests both on the surface of the ground and inside cavities; however, the location and type of protection used for the nest itself can vary from colony to colony. Four of the colonies were located between 323 and 866 metres above sea level, next to banana trees *(Musa paradisíaca)* in a commercial plantation (Fig. 10a- 10b). The nests were located on the ground, covered with a 5cm layer of banana leaves to protect the young (Fig. 10c-lOd). The leaves covering the young were cut into small sizes (<2cm), possibly by the bees. Taylor & Cameron (2003) and González *et al.* (2004) had observed this behaviour in workers of *Bombus transversalis* and *B. atratus* with leaves from other plant species *(Poaceae, Eucalyptus* sp. and *Pinus* sp.). Two other colonies were located more than 800 metres above sea level, also nesting on the surface of the ground, but under a dense cover of grass and with no physical protection other than the grass itself around the cocoons (Fig. lOe-lOf). A seventh colony nested inside the ground, inside a pre-existing cavity 40 cm deep (Fig. lOg-lOh). Only this nest built inside the cavity had a defined entrance to the interior of the colony (5 cm wide and 15 cm long), while in the other colonies the workers entered and left their nests using several different paths.

Some species of *Bombus,* such as *B. lucorum, B. subterraneus* and *B. terrestris,* always nest under the surface of the ground using pre-existing cavities, while *B. hypnorum* prefers to nest in tree hollows (Donovan & Weir, 1978; Goulson, 2003). The Neotropical species *B. atratus, B. morio* and *B. transversalis* nest primarily on the surface of the ground (Dias, 1960; Milliron, 1961; Sakagami *et al,* 1967; Olesen, 1989; Licvano *et al,* 1991; Cameron *et al,* 1999; González *et al,* 2004). However, *B. atratus* is also found in rock crevices (Cameron & Jost, 1998) while *B. brasiliensis* prefers underground nests (Moure & Sakagami, 1962), but *B. pullatus* has been reported nesting in banana leaves, and also in grassy areas, covered by the grass itself (Hines *et al.,* 2007). Therefore, like other *Bombus* species, *B. brevivillus* seems to be opportunistic in terms of nesting sites and the type of protection used for the young.

The adult bees were extremely aggressive, similar to all Neotropical *Bombus* species (Sakagami *et al.,* 1967; Laroca, 1972; Janzen, 1971; Garófalo, 1980; Olesen, 1989; Chavarria, 1996; Cameron & Jost, 1998; Cameron *et al.,* 1999; Garófalo, 2005), and defended the nests fiercely, stinging repeatedly and secreting an unidentified substance into their eyes and onto their skin through their masks. However, this substance did not cause any kind of irritation, the aggressiveness, mainly through stings, was extremely efficient in preventing a long period of observations.

Figure 10. Description of the *Bombus brevivillus* nest in its natural environment in the state of Ceará, Brazil, **(a)** close-up view of the N2 nest showing *B. brevivillus* workers walking outside the nest before leaving to forage; **(b) general** view of the N2 nest built on the ground *and covered with banana leaves cut* into small pieces. *brevivillus workers* walking around the outside of the nest before leaving to forage; (b) general view of the N2 nest built on the ground and covered with banana leaves cut into small pieces; (c) brood area of the N2 nest after partial removal of the banana leaves; **(d)** close-up view of the NI nest showing workers, young (cocoons) and food pots; **(e)** an N4 nest built on the ground, under a dense cover of grass; (f) N4 nest after removal of the grass and the envelope of young; (g) N3 nest after excavation has begun; (h) detail of the queen of the N3 nest at the time of excavation.

* The letter N followed by the number refers to the nest number according to table 5

3.3.2. Description of the *Bombus brevivillus* colony

The architecture of the nest is similar to other *Bombus* species (Olesen, 1989). The brood area containing the larvae of workers, queens and males was located in the centre of the nest, protected by a delicate wax envelope. All the jars are constructed exclusively of wax, and the honey jars were all built along the brood area while the jars containing pollen were found only on the periphery of the nest. The nests occupied an area of 221 to 870 cm and the colonies proved to be populous, ranging from 150 to 502 adult workers and 73 to 149 cocoons (Table 5).

The average number of individuals compared to other Neotropical species was similar to *B. atratus* (Moure & Sakagami, 1962; Benavides, 2008), *B. pullatus* (Janzen, 1971), *B. transversalis* (Cameron & Whitfield, 1996), *B. morio* (Benavides, 2008) and higher than *B. melaleucus* (Hoffinann *et al.,* 2004) and *B. brasiliensis* (Laroca, 1972). The high proportion of brood in relation to adult bees in smaller colonies and a reduction in this proportion in larger colonies shows that young colonies of *B. brevivillus* invest heavily in the production of new brood and in the development of the colony, reaching numbers of more than 500 workers before producing reproductive individuals (Table 5). Highly populated colonies are only common in Neotropical *Bombus* species and this characteristic of *B. brevivillus* can be useful for pollination in large vegetation houses or even in open fields.

Table 5. Characteristics of *Bombus (Thoracobombus) brevivillus* nests sampled in the state of Ceará, Brazil. Only four colonies were captured and have all the information (NI to N4). Data from the colonies (N5 to N7) was limited to location, altitude and nest site

Nest/Coordinates	Honey/pollen pots	Location nest	Area nest (cm$)^2$	Bees adults	Number of cocoons	Ratio of young to adult bees
NI/ S04°29'38.2" W039° 35'42.9"	26	Under banana leaves	864	445	119	0,27
N2/ S04°29'09.0" W039° 35'40.4"	42	Under banana leaves	870	502	149	0,30
N3/ S04° 28'44.6" W039° 34'39.4"	22	Hole in the ground	221	218	78	0,36
N4/S04⁰ 28'45.8" W039° 34' 39.7"	12	Under grass cover	540	150	73	0,48
N5/ S04° 30'53.3" W039° 34'09.7"	-	Under grass cover	-	-	-	-
N6/S04⁰ 30'53.0" W039° 34'15.8"	-	Under banana leaves	-	-	-	-
N7/S04⁰ 30'52.3" W039° 34'13.8"	-	Under banana leaves	-	-	-	-

Figure 11. Nest N2, of the Neotropical species *Bombus (Thoracobombus) brevivillus,* kept in rational breeding conditions in an urban area of Fortaleza, installed in the Bee Sector of the Federal University of Ceará, Brazil.

3.3.3. Foraging patterns, resource collection and the influence of climatic variations on the external activities of workers kept in rational breeding conditions.

The results show that there were significant differences ($P<0.001$) in the flow of bees at the various times of day (Table 6). The highest activity was recorded between 6am and 8am, and these times differed significantly ($P<0.001$) from the others, although they did not differ from each other. From 1pm onwards, the flow of bees and pollen collection did not differ significantly ($P<0.001$) from the other times in the afternoon. There was also no collection of resources from 6pm onwards, when the bees stopped foraging.

With regard to the foraging pattern, even with the colonies installed in rational boxes, the workers were very active in collecting resources, showing the possibility of maintaining *B. brevivilus* colonies in rational breeding conditions. Foraging was concentrated in the morning, and even though the behaviour was quite aggressive, characteristic of *Bombus* species from tropical climates (Garófalo, 1980), the bees allowed their nest to be approached, showing the possibility of using this species in vegetation houses, as is the case with *Bombus terrestris* in various countries around the world, especially in Europe (Velthuis & Doom, 2006).

Table 6. Number of worker bees (means and standard errors) of the native species *Bombus brevivillus* foraging, collecting pollen, nectar/water and ventilating at the nest entrance, under rational breeding conditions. Bee Sector of the Federal University of Ceará, urban area of Fortaleza, Brazil.

Timetable	Flow of bees	Pollen	Nectar/water	Ventilating
05h00	34.03 ± 2.78de	4,12 ± 0,61c	29.86 ± 2.59de	0 ± 0,00e
06h00	75,88 ± 7,73a	14,71 ± 0,98a	61,02 ± 4,08a	0 ± 0,00e
07h00	69,76 ± 4,40a	13,98 ± 1,08a	55,79 ± 3,71a	0 ± 0,00e
08h00	58.64 ± 4.65ab	9,2 ± 0,75b	49.51 ±4.13ab	0,11 ±0,05e
09h00	48.62 ± 4.04bc	5,34 ± 0,54c	43.48 ± 4.73bc	0.29 ± 0.09de
ÍOhOO	36.82 ± 3.34cd	1,3 ±0,18d	35.7 ± 3.30cd	3,58 ± 0,41d
IlhOO	22.79 ± 1.84ef	0,42 ± 0,09d	22.44 ± 1.82e	9.69 ± 1.09bc
12h00	21.62 ± 1.87ef	0,25 ± 0,06d	21.5 ± 1.86def	9,92 ± 1,06b
13h00	18.52 ± 1.60f	0.47 ± 0.1 Od	18.29 ± 1.61efg	14,72 ± 1,41a
14h00	16.2 ± 1.36f	0,11 ±0,03d	16.23 ± 1.35efg	11,17 ± 0,94b
15h00	9,8 ± 0,88f	0,21 ± 0,07d	9.61 ± 0.87fgh	10,8 ± 0,75b
16h00	7,07±0,51f	0,15 ± 0,57d	6.85 ± 0.51gh	7,29 ± 0,86c

| 17h00 | 6,78 ± 0,68f | 0,05 ± 0,02d | 6.79 ± 0.67gh | 0.83 ±0.1 Ide |
| 18h00 | 5,68 ± 0,00f | 0,91 ± 0,00d | 0 ± 0,00h | 0 ± 0,00e |

Values per hour followed by different letters differ significantly from each other (p<0.001)

Table 7. Relationship between day shifts in resource collection and temperature control through ventilation of *Bombus brevivillus* workers kept in rational breeding conditions in an urban area in the municipality of Fortaleza, Ceará, Brazil.

Time of day	Pollen	Nectar/water	Ventilating
Morning	96,23%	84,58%	34,50%
Afternoon	3,77%	15,42%	65,50%

Table 8. Relationship between the flow of bees and pollen collection according to the time of day of *Bombus brevivillus* workers kept in rational breeding conditions in an urban area in the municipality of Fortaleza, Ceará, Brazil.

Time of day	Flow of bees	Bees collecting pollen	%
5-9h	3030	692	22,83%
10-18h	2015	57	2,82%

The collection of food resources (pollen and nectar/water) was concentrated in the morning (Table 7), especially pollen, where practically all the collection took place during this period, with the peak occurring at 06h00 and 07h00 (Table 6). Water/nectar collection was highest between 06:00 and 08:00, decreasing as the flow of workers decreased, however, unlike pollen collection, in the afternoon, from 1pm onwards, a large number of bees were seen foraging, even in the hottest hours of the day (Figure 12).From 1.00 a.m. onwards, the flow of bees, pollen collection and nectar/water collection decreased dramatically (Figure 12), in contrast to ventilation at the nest entrance. No bees were seen thermoregulating the nest until 07:00, and some bees were only seen ventilating (flapping their wings at the colony entrance) from 08:00 onwards, peaking at 13:00.

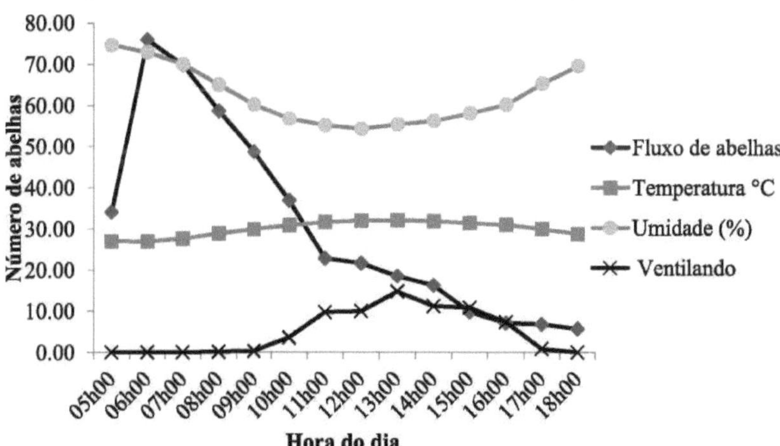

Figure 12. Variations in temperature (°C) and relative humidity (%), the flow of bees and thermoregulation of workers of the native species *Bambus brevivillus* kept under rational breeding conditions in an urban area, in the municipality of Fortaleza, Ceará, Brazil.

The analyses showed a negative Linear Correlation (r) between temperature and bee flow (r = - 0.58399), pollen collection (r = - 0.56221) and necta/water collection (r = - 0.34593), where with increasing temperature the external activities of the workers decreased significantly. The same didn't happen with thermoregulation activity; the results showed a positive correlation (r = 0.6901) between temperature and the number of bees ventilating the entrance to the box (Figures 13 and 14a,b). A Linear Correlation (r) was also carried out between solar radiation and the external activities of the workers, and the results showed the same pattern as that observed for temperature (bee flow (r = - 0.17574); pollen collection (r = - 0.29898); collection of necta/water (r = - 0.07593; bees ventilating (r = - 0.61456), this is logical, since the increase in temperature is directly related to the incidence of sunlight on the land, which is very high in the north-east of Brazil

As for relative humidity, the correlations with bee flow (r = 0.34118) and pollen collection (r = 0.32408) were exactly the opposite of what happened with temperature, i.e. positive correlations, but between humidity and nectar/water collection there was a negative correlation (r = - 0.037495), the same happening with the number of bees ventilating (r = - 0.48018).

In addition to the supply of floral resources, variations in temperature and relative humidity exerted a strong influence on the foraging pattern, resource collection and thermoregulation activities in colonies of *Bombus brevivillus* kept in rational breeding conditions in an urban area in the municipality of Fortaleza, as can be seen in figure 9, from HhOO, when the temperature rises sharply, there is a reduction in resource collection and an increase in the number of bees carrying out thermoregulation (Figure 13).

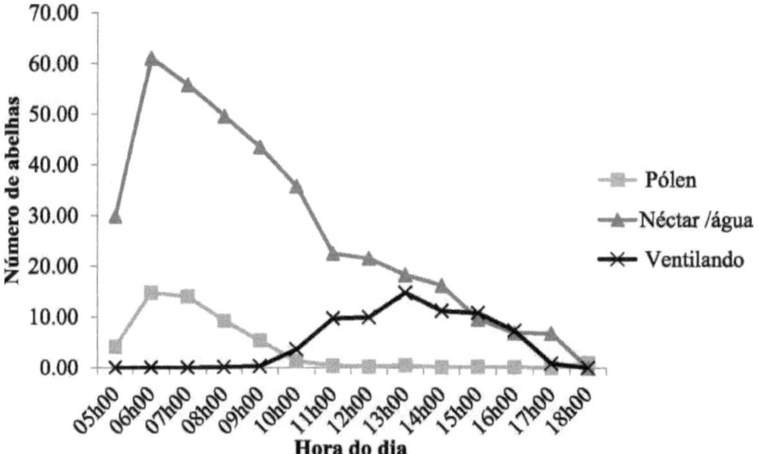

Figure 13. Pollen collection, nectar/water collection and temperature control through ventilation (thermoregulation) of *Bombus brevivillus* workers kept in rational breeding conditions in an urban area in the municipality of Fortaleza, Ceará, Brazil.

The high temperature, especially in the afternoon, which occurs in the north-east region, was an important factor that influenced the workers' external activities and resource collection. The need for thermoregulation had a negative effect on the bees' foraging, as they collected fewer resources and became more agitated and aggressive when entering the rational boxes. Statistical analysis also showed negative correlations between the number of bees ventilating at the entrance to the box (thermoregulation) with the flow of bees (r = - 0.365), pollen collection (r = - 0.31474) and nectar/water collection (r = - 0.097337).

At times when the relative humidity was higher and the temperature lower, there was a greater flow of workers (Figure 12) entering and leaving the nest to forage (bee flow). The collection of resources, pollen and nectar/water, was also greater in these conditions where the temperature was lower and the humidity higher. As the temperature rose and the humidity fell, there was a drop in the number of bees collecting the resources needed to maintain the colony.

However, the number of bees ventilating the nest entrance increased simultaneously with the rise in temperature and the fall in relative humidity. 1pm was the time of day when the highest temperature and lowest humidity were observed, coinciding with the time when the greatest number of bees were ventilating the nest entrance.

Figure 14. **(a)** External movement of *B. brevivillus* workers in the rational box installed in the Bee Sector of the Federal University of Ceará; **(b)** workers flapping their wings and carrying out the thermoregulation process at the entrance to the box; (c) bee foraging and collecting floral resources from the plant species *Tecoma stans* (Bignoniaceae); **(d)** detail of pollen grains in the corbiculae of a worker on returning from the field.

3.3.4. Types of pollen used by the native bee *Bombus brevivillus* under natural conditions

In order to determine the occurrence classes and pollen types used by the *B. brevivillus* bee under natural conditions in Itatira, CE, a total of 4,096 pollen grains were counted. The results showed that the species *Chamaecrista duckeana* (Fabaceae Caesalpinioideae) was predominant in the bees' diet with 3,160 pollen grains (77.14%), the most important pollen type isolated was *Senna splendida* (Fabaceae Caesalpinioideae) with 659 grains and the plant species characterised as occasional isolates, according to the methodology described by Louveaux *et al.* (1978) were *Moringa oleifera* (Moringaceae) with 197 pollen grains and *Momordica charantia* (Curcubitaceae) with 80 grains found in the samples (Table 9).

The municipality of Itatira, in the state of Ceará, is a mountainous region characterised by a great diversity of plant species. However, even with this diversity, the results showed that during part of the year, the Fabaceae family, subfamily Caesalpinioideae, was responsible for 93.22% of the protein diet of the colonies studied in their place of natural occurrence, this family being represented by two different species. The *Chamaecrista duckeana* species was the most representative (Table 9) and is characterised by being a sub-shrub species with medium-sized flowers and poricidal anthers. Pollen is the only resource available to flower visitors, but only a few species of bees adapted to vibration collect pollen from their poricidal anthers, such as bees of the genus *Bombus,* for example (Kiill *et al.,* 2000; Queiroz, 2009).

These results, as already proven for the genus *Bombus,* show that the species *Bombus brevivillus* also has vibratory behaviour and is therefore capable of carrying out pollination by vibration *("Buzz pollinatiorT).* Some agricultural crops, such as tomatoes *(Lycopersicon esculentum),* peppers *(Capsicum annuum* L.) and aubergines *(Solanum malongena),* which have poricidal anthers, have bees of the genus *Bombus* as their main pollinator, *and* these bees improve the productivity and quality of the fruit of these crops (Velthuis & Doom, 2006). These data provide a very encouraging scenario for the use of the native bee *Bombus brevivillus* in the process of pollinating agricultural crops that have poricidal anthers, but more studies are needed to allow for the complete domestication of this bee species and the development of appropriate management techniques for this purpose.

Table 9. Pollen types present in the protein diet of colonies of the native bee *Bombus brevivillus,* under natural conditions of occurrence, in the municipality of Itatira, Ceará, Brazil.

Pollen types	Family	No. of pollen grains	%
Chamaecrista duckeana	Fabaceae	3160	77,14
Senna splendida	Fabaceae	659	16,08
Moringa oleifera	Moringaceae	197	4,80
Momordica charantia	Curcubitaceae	80	1,99
TOTAL	-	**4,096**	**100,00**

3.3.5. Types of pollen used by the native bee *Bombus brevivillus* in rational breeding conditions

The pollen types collected by rationally reared *Bombus brevivillus* workers were identified by counting 2088 pollen grains. The results showed that the predominant plant species (dominant pollen) in the protein diet of colonies in urban areas was *Tabebuia* cf. *impetiginosa* (Bignoniaceae).

The accessory pollen species was *Senna rugosa* (Fabaceae Caesalpinioideae), important isolate *Aniseia martinicensis* (Convolvulaceae) and occasional isolate *Tecoma stans* (Bignoniaceae) (Table 10).

In rational breeding conditions in an urban area in the municipality of Fortaleza, the Bignoniaceae family was an important source of pollen in the diet of *B. brevivilus,* represented by two species and accounting for more than 50 per cent of the pollen collected. The plant species *Senna rugosa* also proved to be an important protein source, confirming the importance of Fabaceae Caesalpinioideae in the protein diet of this bee species.

Table 10. Pollen types present in the protein diet of colonies of the native bee *Bombus brevivillus,* kept in rational breeding conditions in an urban area in the municipality of Fortaleza, Ceará, Brazil.

Pollen types	Family	No. of pollen grains	%
Tabebuia impetiginosa	Bignoniaceae	1126	53,9
Senna rugosa	Fabaceae	665	31,8
Aniseia martinicensis	Convolvulaceae	207	9,9
Tecoma stans	Bignoniaceae	90	4,4
TOTAL	-	**2088**	**100,00**

3.3.6. Development and establishment of colonies and activity of *Bombus brevivillus* queens in rational breeding conditions

Of the four colonies captured and transferred to rational boxes, two did not adapt well to captive conditions, due to attacks by predators (lizards, ants and moths) or problems related to the transfer of the nest. The other two colonies adapted perfectly to these conditions, with normal queen laying, production of immatures (males and females) and intense foraging by the workers.

In *Bombus*, the colony is formed only by the founding queen. During this phase, the queen does all the work of collecting food, building pots and cocoons and caring for the young (Prys-Jones & Corbet, 1991; Delaplane, 1995; Goulson, 2003). When the first worker daughters of this queen begin to emerge, the queen stops her external work and only takes on the reproductive function inside the colony, while the workers take on all the work (Heinrich, 2000).

With the increase in the number of workers and the reduction in the queen's laying rate, the laying of haploid eggs by the workers begins, this moment is known as the *'competition point'* (Duchateau & Velthuis 1988). It is hypothesised that this happens because of the reduction in the queen's pheromone, which is detected by the workers (Cnaani, *et al.,* 2000; Bourke & Ratnieks, 2001).

From this point onwards and after the reproductive individuals (males and new queens) begin to emerge, the colony has a high worker mortality rate, beginning the final part of the reproductive cycle for bees of this genus. With the reduction in the number of workers, food becomes scarce and, consequently, the founding queen dies. At this point, which in temperate climates coincides with the start of the winter season, the sexed workers have already abandoned the colony, which becomes much weaker and more susceptible to attacks from enemies and parasites, leading to the total extinction of the colony at the end of the reproductive cycle (Alford, 1975). However, in tropical climates, according to Sakagami *et al.* (1967) and Garófalo *et al.* (1986) after the death of the founding queen, new queens remain in the nest, and after copulation, reactivate the mother nest, as has been described for the Neotropical species *Bombus atratus.*

Both colonies had a large number of workers, which reduced considerably towards the end of the cycle (Figure 15). However, this large number of workers per colony is a very positive characteristic, making it viable to use the *Bombus brevivilus* species in pollination programmes for agricultural crops.

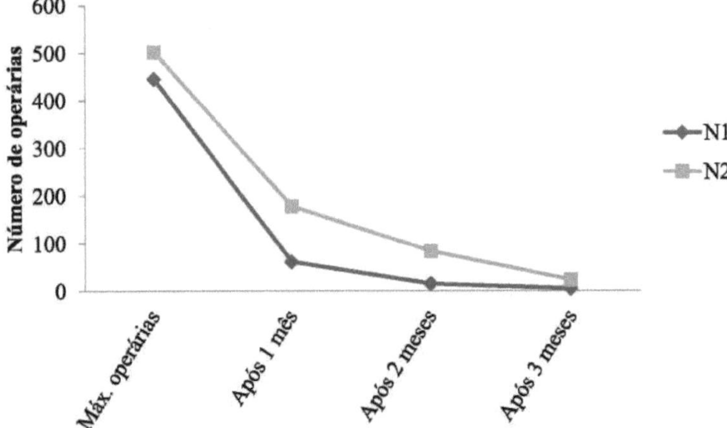

Figure 15   .   Temporal distribution of the number of workers in colonies of *Bombus brevivillus* kept in rational breeding conditions in an urban area in the municipality of Fortaleza, Ceará, Brazil.

With the reduction in the number of workers, food starts to become scarce and, consequently, the founding queen dies. At this point, which in temperate climates coincides with the start of the winter season, the sexes have already abandoned the colony, which weakens considerably and becomes more susceptible to attacks from enemies and parasites, leading to the total extinction of the colony at the end of the reproductive cycle (Alford, 1975).

The queens from both colonies laid normally after being transferred to the rational boxes and were constantly walking on the cocoons (Figure 16a), showing a lot of activity. At the time of the transfers, both nests (NI and N2) already had a large number of male and queen cocoons. This identification was possible because the cocoons of the reproductive individuals are much larger than those of the workers. A few days after the colonies were set up in the Bee Sector, the first males and queens began to emerge. They remained inside the box for four to six days before leaving the colony. The number of males produced was much higher than the number of queens (Table 11) and in colony N2, during the emergence of the sexes, the old queen tried to maintain dominance of the nest by laying new eggs (Figure 16d), a behaviour not observed in colony N2.

Table 11. Number of reproductive individuals and the ratio of males to queens in two colonies of the native species *Bombus brevivillus* kept under rational breeding conditions in an urban area in the municipality of Fortaleza, Ceará, Brazil.

Nest	Males	Queens	males : queen
NI	80	17	4,7: 1
N2	98	15	6,5 : 1
TOTAL	**178**	**32**	-

In all, 210 individuals were produced, including males and queens, in just two colonies. This characteristic opens up a vast field for future research into the production of colonies of the native species *Bombus brevivillus* in the laboratory, as has been done in Europe with other *Bombus* species for a long time. This would be a major advance for commercial agricultural pollination programmes, as well as contributing to the preservation of the species by multiplying new nests and introducing them into regions where their presence has been reduced by human activity.

It is therefore extremely important to master and develop artificial mating techniques that enable the production of several colonies. This model already exists in Europe and serves as a benchmark, where several specialised companies produce thousands of colonies every year.

According to Garófalo (1980), colonies of Neotropical *Bombus*, unlike those of temperate species, can be perennial, meaning that the nest does not die out at the end of the biological cycle. A daughter queen mates and takes the place of the mother queen, reactivating the colony in the same place as the previous generation. One of the colonies (N2), after the reduction in the number of workers and the production of sexuates, went completely extinct, with only the queen and a few workers remaining before the colony went completely extinct (Figure 16h). Therefore, it showed a biological cycle that is quite characteristic of the genus, and similar to temperate climate species.

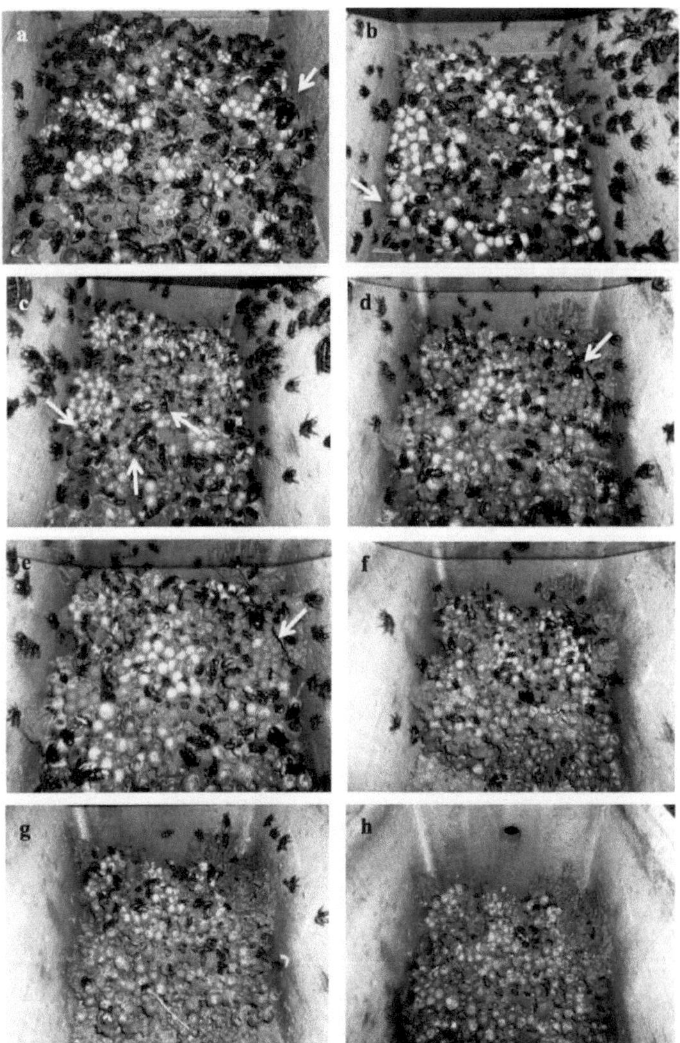

Figure16   .   Colony (N2) of *B. brevivillus* kept in rational breeding conditions, in an urban area, in the municipality of Fortaleza, Ceará, Brazil, **(a)** Large number of workers and detail of the presence of the queen on the cocoons, at the time of capture the colonies had already started laying immatures; **(b)** colony well established in the rational box, workers foraging normally, normal laying of the queen, detail of the cocoons of reproductive individuals becoming clearer and about to hatch; **(c)** emergence of the princesses; **(d)** queen trying to maintain dominance by increasing her laying rate; **(e)** close-up of the queen's new postures; (f) drastic reduction in the number of bees; (g) all the males and princesses have left the colony, leaving only the queen and a few younger workers **(h)** moment before the colony's total extinction, leaving only the queen and five workers.

Species from temperate climates, with extreme temperatures in harsh winters, exhibit hibernation behaviour. Thus, in temperate *Bombus* species, the biological cycle is well defined, with the colony dying out after the production of reproductive individuals, which leave the colony to found new nests elsewhere.

However, this same pattern did not occur with one of the colonies of the native species *Bombus brevivillus* studied, colony (Nl). During the period of colony weakening and reduction in the number of individuals, but before the colony became extinct, a daughter queen took over the laying and reactivated the colony as we can see in figure 17g and 17h. Due to an attack by moths, the new bees produced by this new queen left the rational box and it was not possible to obtain any further information about this colony.

However, despite the low number of repetitions (only two colonies), the daughter queen was observed laying and starting a new nest in the same place as the mother colony, showing the possibility of nest reactivation in the native Neotropical species *Bombus brevivillus*. This behaviour has never been described so directly in previous research and serves as an incentive for further conservation research into this native species.

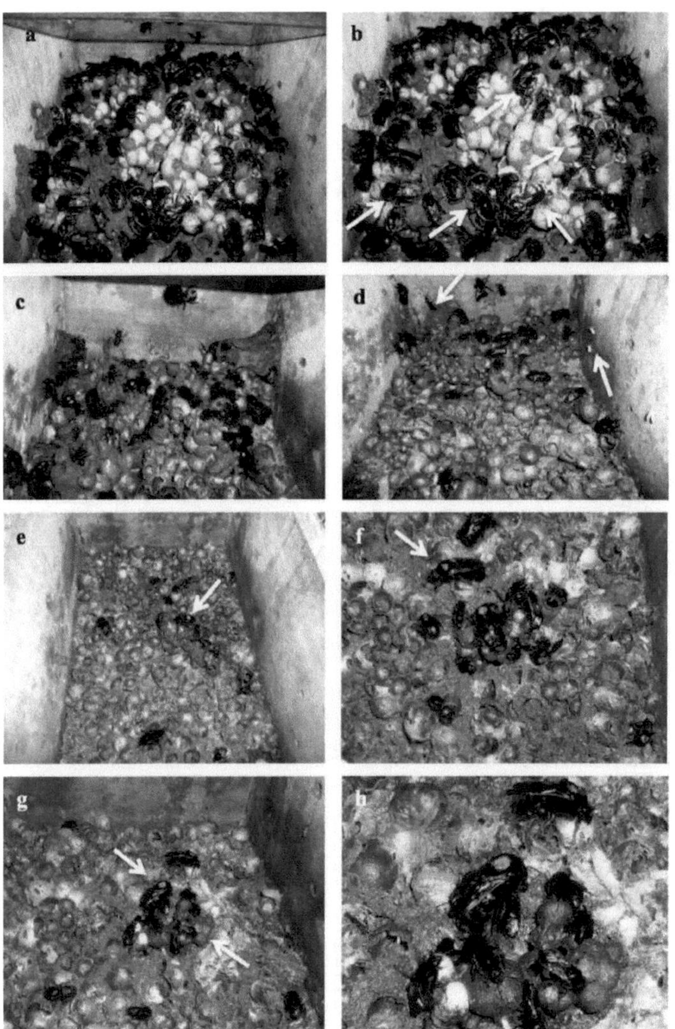

Figure17 . __ Colony (Nl) of *B. brevivillus* kept in rational breeding conditions in an urban area in the municipality of Fortaleza, Ceará, Brazil, **(a)** Production of males and princesses; **(b)** close-up of the large number of reproductive individuals **(c)** drastic reduction in the number of bees; **(d)** before the colony was reactivated, still with some princesses, workers and the presence of natural enemies; **(e)** princess starting to work on forming the new colony, queen no longer present; (f) start of laying, detail of a princess from another colony; (g) new queen and recent laying, with the presence of newly hatched workers helping to form and reactivate the new colony; **(h)** close-up of the new *B.* brevivillus nest. *brevivillus.*

3.4. CONCLUSIONS

It can be concluded that *Bombus brevivillus* seems to be opportunistic in terms of nesting sites and the type of protection used for the young, not investing much effort in building the nest.

The architecture of the nest is similar to that of other *Bombus* species. Colonies of *B. brevivillus* can be useful for pollination in large vegetation houses or even in open fields, as they invest a lot in the production of new young and in the development of the colony.

The Fabaceae and Bignoniaceae families have important species for the diet of *B. brevivillus,* serving as food sources for the colonies, both in natural conditions and in rational breeding conditions.

Even under captive conditions, *B. brevivilus* workers foraged normally and produced males and queens, opening up a range of opportunities for their rational breeding and for the production of colonies in the laboratory, for conservation purposes and commercial pollination programmes.

Unlike temperate bees, *Bombus brevivillus* workers foraged intensively in the early morning, when the temperature is lower and the relative humidity higher, especially to collect pollen. The collection of water and/or nectar was distributed throughout the day, and the need to thermoregulate in the hottest hours had a negative influence on the collection of food resources in the field.

There is the possibility of maintaining colonies of *B. brevivilus* in rational breeding conditions, with the development of management techniques. The males and queens obtained through this system could be used to produce new colonies in the laboratory.

3.5. BIBLIOGRAPHICAL REFERENCES

BARTH, O.M. Pollen in monofloral honeys from Brazil. **Journal of Apicultural Research,** v. 29, p. 89-94,1990.

BENAVIDES, M.L.A. **Aspects of the reproductive biology of *Bombus morio* (SWEDERUS) and *Bombus atratus* FRANKLIN (HYMENOPTERA, APIDAE).** Master's dissertation. Master's degree in entomology. Federal University of Viçosa, Minas Gerais. 2008.

CAMERON, S.A. & JOST, M.C. Correlates of dominance status among queens in a cyclically polygynous Neotropical bumble bee. **Insectes Soc,** v. 45, p. 135-149, 1998.

CAMERON, S.A & WHITFIELD, J.B. Use of walking trails by bees. **Nature,** v. 379, p. 125,1996.

CAMERON, S.A.; WHITFIELD, J.B.; COHEN, M. & THORP, N. **Novel use of walking trails by** the Amazonian bumble bee, *Bombus transversalis* **(Hymenoptera: Apidae)**, in: BYERS G.W., HAGEN R.H., BROOKS R.W. (Eds.), Entomological contributions in memory of Byron A. Alexander, Univ. Kans. Nat. Hist. Mus. Spec. Publ. 24, pp. 187-193, 1999.

CARVALHO-ZILZE, G.; PORTO, E.L.; SILVA, C.G.N. & PINTO, M.F.C. Flight activity of *Melipona seminigra* (Hymenoptera: Apidae) workers in an Amazonian agroforestry system. **Bioscience Journal,** v. 23, p. 94-99, 2007.

COUTO, R. H. N. & COUTO, L. A. **Apicultura: manejo e produtos.** 2 ed. Jaboticabal: FUNEP, 191 p., 2002.

CRUZ, D. O.; FREITAS, B. M.; SILVA, L. A.; SILVA, E. M. S. & BOMFIM, I. A. Pollination efficiency of the stingless bee *Melipona subnitida* on greenhouse sweet pepper *(Capsicum annuurn).* **Revista Agropecuária Brasileira,** v. 40, p. 1197-1201, 2005.

CHAVARRIA, G. Notes on a combined nest of *Bombus pullatus* (Hymenoptera: Apidae) and *Acromyrmex octospinosus* (Hymenoptera: Formicidae). **Journal Kansas Entomological Society,** v. 69, p. 403-405, 1996.

DIAS, D. Contribuição para o conhecimento da bionomia de *Bombus incarum* Franklin da Amazônia (Hymenoptera, Bombidae). **Revista Brasileira de Entomologia,** v. 8, p. 1-20, 1958.

DIAS, D. Note on a *Bombus* nest built above ground (Hymenoptera, Apoidea). **Revista Brasileira de Entomologia,** v. 9, p. 151-156,1960.

DONOVAN, B.J. & WIER, S.S. Development of hives for field population increase, and studies on the life cycle of the four species of introduced bumble bees in New Zealand. **N Z J Agricutural Research,** v. 21, p. 733-756, 1978.

FAO. Conservation and management of pollinators for sustainable agriculture - the International response. In: Freitas, B.M.; Pereira, J.O.P. (eds.) Solitary bees: conservation, rearing and management for pollination. University Press.
Fortaleza, Brazil, p. 19-2, 2004.

FIUME, F. & PARISI, B. Fitoregolatori e bombidi nella frittificazione del pomodoro. **Colture**

Protette. 10: 87-93,1994.

FREITAS, B.M. Use of rational pollination programmes in agricultural areas. **Messagem Doce,** São Paulo, v. 46, p. 16-20,1998.

FREITAS, B.M.; IMPERATRIZ-FONSECA, V.L.; MEDINA, L.M.; KLEINERT, A.M.P.; GALETTO, L.; NATES-PARRA, G. & QUEZADA-EUÁN, J.J.G. Diversity, threats and conservation of native bees in the Neotropics. **Apidologie,** v. 40, p. 332-346, 2009.

GAROFALO, C.A. Bombus: the ground-dwelling mamangavas and their importance as pollinating agents, 2005. Publishing online.
http://www.apacame.org.br/mensagemdoce/80/msg80.htm. Accessed 11 May 12

GARÓFALO, C.A. Bionomic aspects of *Bombus* (Fervidobombus) *morio* (Swederus). III. Worker size and colony development (Hymenoptera, Apidae). **Revista Brasileira de Biologia,** v. 40, p. 345-348,1980.

GONZÁLEZ, V.H.; MEJIA, A.; RASMUSSEN, C. Ecology and nesting behaviour of *Bombus atratus* Franklin in Andean highlands (Hymenoptera: Apidae). **Journal Hymenoptera Research,** v. 13, p. 234-242, 2004

GOULSON, D. **Bumblebees: Behaviour and Ecology.** Oxford UK: Oxford Univ. Press, 2003.

GRANDE, L.; LUZ, J.M.Q.; MELO, B.; LANA, R.M.Q. & CARVALHO, J.O.M. The protected cultivation of vegetables in Uberlândia-MG. **Horticultura Brasileira,** v. 21, n. 2, p. 241-244, 2003.

HINES, H.M.; CAMERON, S.A. & DEANS, A.R. Nest architecture and foraging behaviour on *Bombus pullatus* (Hymenoptera: Apidae), with comparisons to other tropical bumble bees. **Journal Kansas Entomological Society,** v. 80, p. 1-15, 2007.

HOFFMANN, W.R.E.; TORRES, A. & NEUMANN, P. A scientific note on the nest and colony development of the Neotropical bumble bee *Bombus (Robustobombus) melaleucus.* **Apidologie,** v. 35, p. 449-450, 2004.

IPECE. Basic municipal profile, 2012: Fortaleza.

http://www.ipece.ce.gov.br/publicacoes/perfil_basico/pbm-2012/Fortaleza.pdf.
Accessed 20 Aug 14.

IPECE. Basic municipal profile, 2012: Itatira.
http://www.ipece.ce.gov.br/publicacoes/perfil_basico/pbm-2012/Itatira.pdf. Accessed 27 Jun 14.

JANZEN, D.H. The ecological significance of an arboreal nest of *Bombus pullatus* in Costa Rica. **Journal Kansas Entomological Society,** v. 44, p.210-216, 1971.

KEVAN, P.G. & PHILLIPS, T.P. The economic impacts of pollinator declines: an approach to assessing the consequences. **Conservation Ecology,** v. 5, p. 8, 2001.

KIILL, L.H.P.; HAJI, F.N.P. & LIMA, P.C.F. Floral visitors of invasive plants in irrigated fruit areas. **Scientia Agrícola,** v. 57, p. 575-580, 2000.

KLEIJN, D. & RAEMAKERS, A. A retrospective analysis of pollen host plant use by stable and declining bumblebee species. **Ecology,** v. 89, p. 1811-1823, by the Ecological Society of America, 2008.

KLEIN, A.M.; VAISSIERE, B.E.; CANE, J.H.; STEFFAN-DEWENTER, L; CUNNINGHAM, S.A.; KREMEN, C. & TSCHARNTKE, T. Importance of pollinators in changing landscapes for world crops. **Proceedings of the Royal Society B: Biological Sciences,** v. 274, p. 303-313, 2007.

LAROCA, S. On the bionomics of *Bombus brasiliensis* (Hymenoptera, Apoidea). **Acta Biológica,** v. 1, p. 7-28, 1972.

LAROCA, S. On the bionomics of *Bombus morio* (Hymenoptera, Apoidea). **Acta Biológica,** v. 5, p. 107 - 127, 1976.

LIEVANO, A.; OSPINA, R. & NATES, G. Altitudinal distribution of the genus *Bombus* in Colombia (Hymenoptera: Apidae). **Invertebrados,** v. 4, p. 541-550,1991.

LIMA-VERDE, L.W. **Melissofaunal resources of the Baturité Massif, Ceará, Brazil - diversity and zootechnical potential.** Doctoral thesis. PhD in Zootechnics, Federal University of Ceará,

2011.

LOUVEAUX, J.; MAURIZIO, A. & VORWOHL, G. Methods of Melissopalynology. **Bee World,** v. 59, p. 139-157, 1978.

MAGALHÃES, C.B. & FREITAS, B.M. Introducing nests of the oil-collecting bee *Centris analis* (Hymenoptera: Apidae: Centridini) for pollination of acerola *(Malpighia emarginatd)* increases yield. **Apidologie,** v. 44, p. 234-239, 2013.

MALAGODI-BRAGA, K.S. & KLEINERT, A.M.P. Could *Tetragonisca angustula* (Apinae, Meliponini) be effective as strawberry pollintor in greenhouse. **Australian Journal of Agriculture! Research,** v. 55, p. 771-773, 2004.

MILLIRON, H.E. Notes on the nesting of *Bombus morio* (Swederus) (Hymenoptera, Apidae). **Can Entomol,** v. 93, p. 1017-1019, 1961.

MOORE, P.D. & WEBB, J.A. **Anales illustrated guide to pollen analysis.** London: Hodder and Stoughton, 133p., 1978.

MOURE, J.S. & MELO, G.A.R. **Catalogue of Bees (Hymenoptera, Apoidea) in the Neotropical Region,** 2012. Online, http://www.moure.cria.org.br/catalogue. Accessed 22 Oct 13.

MOURE, J.S. & SAKAGAMI, S.F. The social mamangabas of Brazil *(Bombus* Latreille) (Hymenoptera, Apoidea). **Study Entomological,** v. 5, p. 65-194,1962.

OLESEN, J.M. Behavior and nest structure of the Amazonian *Bombus transversalis* in Ecuador. **Journal Tropical Ecology,** v. 5, p. 243-246,1989.

QUEIROZ, L.P. **Leguminosas da caatinga.** Feira de Santana: State University of Feira de Santana, 467p., 2009.

SAKAGAMI, S.F.; AKAHIRA, Y. & ZUCCHI, R. Nest architecture and brood development in a neotropical bumblebee *Bombus atratus.* **Insectes Society,** v. 14, p. 389^414, 1967.

SANDE, J. van der. Bumblebees are a good altemative to truss vibration for beefsteak tomatoes.

Hort. Abstr., v.60, n. 506, 1990.

SARAIVA, A.M.; ACOSTA, A.L.; GIANNINI, T.C.; IMPERATRIZ-FONSECA, V.L & MARCO JÚNIOR, P. *Bombus terrestris* in South America: Possible invasion routes of this exotic pollinator to Brazil. In: **Pollinators in Brazil: contributions and perspectives for biodiversity, sustainable use, conservation and environmental services.** IMPERATRIZ-FONSECA, V.L.; CANHOS, D.A.L.; ALVES, D.A.; SARAIVA, A.M. (eds.), São Paulo: Edusp, chap. 10, p. 203-212, 2012.

SILVA-MATOS, E.V. **Studies on size variability among workers of *Bombus atratus* (Hymenoptera, Apidae).** Dissertation, University of São Paulo, 1986.

TAYLOR, O.M. & CAMERON, S.A. Nest construction and architecture of the Amazonian bumble bee (Hymenoptera: Apidae). **Apidologie,** v. 34, p. 321-331,2003.

VELTHUIS, H.H.W. & DOORN, A. van. A century of advances in bumblebee domestication and the economic and environmental aspects of its commercialisation for pollination. **Apidologie,** v. 37, p. 421-451, 2006.

WILLIAMS, P.H.; CAMERON, A.S.; HINES, H.M.; CEDERBERG, B. & RASMONT, P. A simplified subgeneric classification of the bumblebees (genus *Bombus).* **Apidologie,** v. 39, p. 46-74, 2008.

WODEHOUSE, R.P. **Pollen Grains,** Mac Graw-Hill Co., New York, 574p., 1935.

ZUCCHI, R. **Bionomic aspects of *Exomalopsis aureopilosa* and *Bombus atratus* including considerations on the evolution of social behaviour (Hymenoptera, Apoidea).** Doctoral thesis, University of São Paulo, 1973.

FINAL CONSIDERATIONS

Research around the world has been warning for a long time about the decline in pollinator populations, as well as drastic reductions in the abundance and diversity of plant and animal species, especially since the second half of the last century.

The constant and growing destruction of the environment reduces the availability of food in nature and bees, because they depend solely and exclusively on floral resources, are one of the groups of animals particularly affected.

It is therefore extremely important to carry out further research into bees, which will help advance specific knowledge of each species and create conservation programmes.

There also needs to be a change of mentality on the part of everyone. Before it's too late, there will have to be more initiatives for the preservation of the environment in order to reverse this degrading and advanced picture of total environmental destruction.

Buy your books fast and straightforward online - at one of world's fastest growing online book stores! Environmentally sound due to Print-on-Demand technologies.

Buy your books online at
www.morebooks.shop

Kaufen Sie Ihre Bücher schnell und unkompliziert online – auf einer der am schnellsten wachsenden Buchhandelsplattformen weltweit! Dank Print-On-Demand umwelt- und ressourcenschonend produziert.

Bücher schneller online kaufen
www.morebooks.shop

Printed by Books on Demand GmbH, Norderstedt / Germany